MAINTAINING FISHES
for Experimental and Instructional Purposes

MAINTAINING FISHES

for Experimental and Instructional Purposes

William M. Lewis

Southern Illinois University Press · Carbondale

PREFACE

THE USE OF FISHES as experimental animals has contributed much to our knowledge in many phases of animal science. Nigrelli (1953) has reviewed some of the more outstanding of these contributions. The principal obstacle connected with the use of fishes as experimental animals is the lack of information on the selection and maintenance of experimental fishes. This apparent need has prompted the author to attempt to bring together some of the more important considerations associated with this problem. Notable among these considerations are nutrition, disease, and the artificial maintenance of suitable environmental conditions.

This work is limited to a treatment of freshwater fishes and is based upon the author's experience in this area of interest and upon such literature as seems relevant. There are some considerations concerning the maintenance of captive fishes about which there is little information available. It is hoped that research persons will be stimulated to make contributions on these topics.

The author wishes to thank his wife, Sue D. Lewis, and Mr. Vernon Cole and Mr. Robert Summerfelt of the Southern Illinois University Fisheries Research Laboratory for contributing to this work.

William M. Lewis

June 28, 1962

CONTENTS

ILLUSTRATIONS

MAINTAINING FISHES
for Experimental and Instructional Purposes

A Classification of the Biological States of Waters and of Fish-holding Procedures

FISHES ALTER the water in which they live. They remove dissolved oxygen and contaminate the water with fecal, respiratory, and urinary wastes. When fishes are fed nonliving food, the water is further contaminated by soluble components of the food as well as any uneaten portion of nonsoluble constituents. There is, of course, a direct relationship between the degree of crowding of the fishes and the extent to which the water is contaminated or altered.

The different contaminants affect the water in different ways. Fecal and unutilized food materials cause the water to become turbid, cause a build-up of carbon dioxide and ammonia, and increase the number of decay organisms, especially bacteria. If unusually abundant, these organisms can seriously reduce the dissolved oxygen content. Some of the saprophytes such as the water molds *Saprolegnia spp.* may become parasitic on the fishes or their eggs.

The presence of fishes increases the dissolved salt content of water and alters the composition of its ionic system. The effect of this change on fishes is not known. Short of pollution, fishes alter water in such a manner as results in increased growth of other fishes subsequently occupying it. This phenonemon is known as conditioning.

Fishes can be crowded to the extent that both reproduction and growth are inhibited. Exactly what these inhibiting substances are is not known. The situation may be nothing more than a combination of waste materials concentrated enough to be generally toxic to fishes and thus interfere with all normal functions.

Natural processes eliminate or inactivate some waste materials. Green plants, including algae, take up large quantities of carbon dioxide and release oxygen. Bacteria break down fecal and unused food materials. Snails and other invertebrates utilize excess food materials. Chemical reactions cause the breakdown and inactivation of many wastes. Still other wastes are lost as gases at the water's surface. But none of the natural processes are entirely adequate when fishes are crowded beyond a certain point. We do not presently know the range of density values which constitute the upper limit that can be accommodated by natural processes. In terms of grams of fish per liter of water it is probably less than one gram per liter in the absence of an algal bloom or other plant growth. Higher densities appear possible when plant growth is prevalent. The relationship will vary with species of fish and a number of other considerations.

Among tropical fish hobbiests the rule of thumb for

calculating desirable densities is an inch of fish per gallon of water. Gordon (1950) calls attention to a recent trend toward using a ratio between surface area and inches of fish. The latter method does have the value of emphasizing the importance of surface area in the exchange of gases between the water and atmosphere. However, in aerated tanks surface exchange might not be as important as volume. The author is of the opinion that density should be specified in terms of grams of fish per liter of water, and, where flushing is done, the rate of flushing should be specified in terms of percent change of a tank's volume in twenty-four hours.

In order to obtain a clearer understanding of these various alterations of water by fishes we will classify waters as to their biological states using the categories: new, aged, conditioned and polluted. Particular attention is called to the need for making a distinction between aged and conditioned waters. It will be noted that there is inadequate information on conditioning of water and the polluting of water by fishes. Nevertheless, we can profit by making use of available information.

The Biological States of Water

New water is water that has not supported fish life and is essentially free of other biota. There are at least two possible sources of such water: spring or well water and rain water collected in an earthen pool. Aged or conditioned water (see below) or even some types of polluted water may be boiled or otherwise treated to give it the characteristics of new water.

Aged water we will define as that which has not

been utilized by fishes but has been permitted to stand long enough to develop a population of microscopic and semimicroscopic plants and animals. Sunlight and the addition of a small amount of organic matter result in more rapid and complete aging. In addition to involving the development of microscopic life, aging also permits the loss of excess quantities of gases and the precipitation of toxic minerals sometimes present in new water, especially water from wells and springs.

Conditioned water is water which has supported fish life. It has been demonstrated that at least under some conditions fish grow better in conditioned water. Most aquarists already recognize the importance of this phenomenon for it is a common practice to add at least some "old water" to a newly established aquarium. It is not yet known what is involved in conditioning. Perhaps the best theory to date has been offered by Breder (1931). He suggests that water previously occupied by fishes contains a bacteriophage that controls the growth of undesirable bacteria. Other possible explanations of conditioning might include changes in the ionic system of the water brought about by exchanges of salts between the fishes and the water, or we might say that in the presence of fishes the ionic system of the water is adjusted to more nearly suit that of the fishes' requirements. Also relating to the ionic system is the possibility that toxic ions are absorbed by the first fishes and thus leave the water more favorable for the subsequent inhabitants.

Under the heading of "Polluted Water" it is necessary to recognize a number of subdivisions. It will be evident that some of the pollutants are produced by the fishes occupying the water while others are from outside sources.

Any materials which will readily decay we may class as putrescible. Associated with the decay of such materials is the utilization of the available dissolved oxygen and the production of carbon dioxide and ammonia. The best example of this class of pollutants is excess food. Fecal material is also putrescible and has been shown by Kawamoto (1961) to result in a serious build-up of ammonia and a reduction in dissolved oxygen.

Ammonia is considered to be a highly toxic material, and the accumulation of it in aquarium water is undesirable. In addition to the ammonia produced from putrescible materials ammonia is excreted by fishes. Urea is also excreted and breaks down to form additional ammonia. It is thus evident that ammonia is a principal pollutant of aquarium water, and aquarium management should involve arrangements for disposing of it.

Under an aquarium management system where the water level is maintained by the addition of water without any draining of the old water there is a gradual increase in the salt content of the water. In addition, there is probably on the part of the fishes some selective removal of ions and an addition of others to the water. It is questionable if this abnormal condition of the salt content is desirable.

When some fishes are crowded they cease to reproduce. Since water utilized by a crowded fish population will reduce reproduction in an uncrowded population, it appears that fishes produce some substance that represses reproduction (Swingle, 1956). We may thus consider this material as a water pollutant. This phenomenon is well enough recognized that in hatcheries it is utilized as a means of preventing goldfish from spawning too early in the spring. The fish are crowded during the early spring and moved to a less crowded condition when the operator desires them to commence spawning.

It has been shown (Rose, 1959) that tadpoles release into the water a material which inhibits the growth of other tadpoles, and there is some evidence that such a material may be produced by fishes. If such is the case, the matter should prove to be of great interest to aquarists. In order to illustrate the possible nature of such substances we might review some of the characteristics of the material that Rose worked with in tadpoles.

When tadpoles are crowded, one or two of a group will greatly exceed the others in growth.

The larger tadpoles produced something that inhibited the growth of the smaller ones.

If small tadpoles in a separate tank were exposed to water from a tank containing large tadpoles the growth of the smaller ones was inhibited.

Inhibiting is prevented by heating the water to 60° to 70° C., by filtering through fine filter paper or by aging the water for a month.

The presence of plants and other animals reduces the inhibiting effect.

The inhibitive agent is apparently produced only by rapidly growing tadpoles. It is thus said to be a feed-back phenomenon.

The inhibiting agents are specific although inhibition between members of the same genus appears to occur.

Both the number produced and survival of guppies has been shown to be controlled by an inhibitor.

We might now make some generalities on aquarium management that take into consideration the foregoing information. New water should be used for flushing aquaria and for making aged and conditioned water. Aged water is ideal as an environment for young fishes of all species. In the culture of many fishes aged water for fingerlings is considered a necessity. For small species of fishes that utilize the tiny plants and animals contained in aged water, aged water would be of value for flushing tanks.

Conditioned water is a matter of concern in growth studies. In this type study the initial filling of the aquaria should be with conditioned water. It is possible that only a part of the water need be conditioned, i.e., a tank might be filled with new or aged water and then a portion of conditioned water added.

The primary concern with polluted water is the development of aquarium management procedures that avoid the accumulation of putrescible materials. As pointed out above, food and fecal material are normally the principal source of trouble. Fecal material should be removed by periodically siphoning off the bottom of the aquaria. Trouble from food is avoided by not overfeeding and by the selection of a food that is not too rapidly soluble in water. Some foods, as for example, ground liver, are very troublesome to feed, while live foods such as water fleas (*Daphnia*) and brine shrimp (*Artemia*) seldom produce a pollution problem. Ammonia and other gases are produced by decay and as waste products from fishes. These can be prevented from accumulating by aeration and charcoal filtration.

The accumulation of soluble salts both by evaporation and from excretion is best avoided by frequently replacing a portion of the water of each tank or by continuously flushing the tanks.

The repression of reproduction appears to become evident when populations are unusually dense. Thinning a population is an obvious approach. Continuous flushing may also be a solution.

Growth repression may function at a much lower level of population density than reproduction repression. The author does not feel that enough is known about this phenomenon to make possible recommendations for avoiding it.

Classification of Fish-holding Facilities

In order to increase the grams of fish per liter of water, the following artificial aids are commonly used: frequent water changes, aeration, filtering and flushing. Whether or not artificial aids are used and to what extent will considerably affect the physicochemical conditions of the aquarium water. The variation in artificial aids in maintaining desirable aquarium conditions must be recognized in deciding what type of set-up is best suited to a particular need. We, thus, may consider the different arrangements one might set up and the use to which they are best suited.

The first and simplest arrangement involves no replacement or artificial aids for neutralizing wastes and replenishing oxygen. The weight of fish to volume of water must be low enough to permit natural processes to maintain the water in a condition satisfactory for fishes. Higher aquatic plants are often added to this set-up. They increase the available oxygen and decrease the carbon dioxide, at least when they are exposed to light, and probably have the over-all effect of making it possible to hold more fish per volume of water. This arrangement most closely resembles natural pond conditions and is the best choice of arrangements when one seeks a good choice for display tanks. It requires less care, may be made to look quite natural and does not require a water supply or drain service.

The second arrangement includes a partial change of water daily and continuous aeration. This type of arrangement permits holding a considerable weight of fish in a limited space and without much special equipment. It is of value for testing toxicity of chemicals and for the study of rapidly developing diseases such as those caused by bacteria and some protozoans.

In the third basic type set-up the water is circulated and filtered. This type of holding facility is of special value where one wishes to maintain dense populations of fish over an extended period of time and there is a scarcity of good water of the correct temperature.

The fourth arrangement involves continuous replacement of the water by flushing. The amount of flushing may vary considerably. Holding tanks in goldfish hatcheries may have a flush rate high enough to give the water a change every twenty to thirty minutes, but for aquaria where the fish are much less crowded, a flush rate giving a complete change in twenty-four hours is sufficient to maintain clean water. Of course, more flushing permits the holding of more fish per volume of water.

If experimental conditions and facilities permit, an arrangement involving even a minimum amount of flushing is the best choice for maintaining fishes in captivity. Problems of accumulated, dissolved wastes and the concentration of salts by evaporation, are corrected by continuous flushing.

The Aquarium Building

IT IS OFTEN assumed that a greenhouse might constitute an ideal aquarium building. The author is not in agreement with this view. An aquarium building does not require nearly as much light as is afforded by a greenhouse. In fact, the excess light may constitute a problem. Especially where glass aquaria are used, excess light encourages algal growth and makes it difficult to keep tanks reasonably clean. Although large concrete or metal tanks can tolerate considerable light without developing algal problems, for general use, the light problem in a greenhouse makes it undesirable as an aquarium building. In addition, a greenhouse has a very high heat loss and is usually more expensive to construct than a more conventional-type building.

The author recommends a building of the following general design:

CONSTRUCTION: Brick veneer or frame.

SIZE: 15 feet wide, 30 or more feet in length.

ROOF: Saddle type with an overhang of only a few inches.

WINDOWS: Row of windows along entire extent of side walls. Windows to be 2 feet in height and placed 6 feet above floor level.

FLOOR: Poured concrete with drains.

FOUNDATION: Extending eight inches above floor level.

INSULATION: Below floor, in side walls and overhead.

Heating

Adequate heating is one of the more important considerations in the equipping of an aquarium building. Certain peculiarities of aquarium buildings affect the type and design of the heating systems. Higher than normal temperatures are often required. Thus, a building devoted primarily to tropical fishes should be maintained at 80 degrees F. Ideal lighting and ventilation of an aquarium building is made possible by a building design with an abnormal extent of outside walls. Such a building even though well insulated will have a high heat loss. This characteristic also makes it difficult to maintain an even distribution of heat unless a special effort is made to avoid this problem at the time of installation of the heating system. A final consideration is the frequent need to maintain different temperatures in different tanks.

A hot water heating system is probably more satisfactory for smaller buildings while steam heat would be most satisfactory for larger installations. A hot air system would be a very poor choice. Whether steam or hot water heat is chosen, the heat radiating units should extend along all walls and should include zone control.

Electrical Service

The electrical service to an aquarium room should be of high capacity. As an approximation each 400 square feet of floor space should have an independent, 110 volt, number 10 wire circuit. Convenience outlets should be placed approximately 5½ feet from the floor (See Figure 1 for location various services). Wiring should be in conduit and special attention should be given to proper grounding. All outlets should accommodate a grounding wire from the appliances. Tank stands, air lines, and all other metal objects should be thoroughly grounded. The danger of electrical shock in an aquarium room is obvious.

Lighting

Information on the light requirements of fishes is limited. We know little or nothing about their quantitative or qualitative needs. They, of course, require a minimum amount of light for vision and may utilize light in the synthesis of certain vitamins. In regard to the latter, however, a vitamin fortified diet would constitute a suitable substitute. There are several good

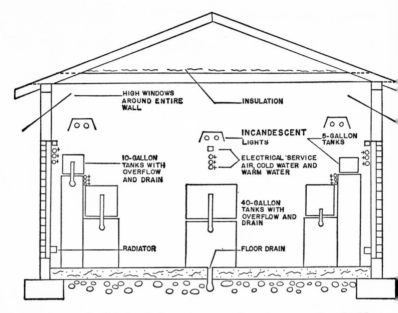

1. *Sectional view of a general purpose aquarium building.*

works relevant to fishes' response to length of illumination. Some fishes show marked photoperiodicity in their spawning behavior.

The importance of light to plants is readily observed, however, and the intensity of illumination is frequently based more on the requirements of aquarium plants and the requirements for the control of algal growth than upon the possible needs of the fishes. The recommendations on lighting that follow are based largely on supposition and experience.

Natural lighting for the aquarium building is furnished by the windows suggested in the building plan. Gordon (1950) rather strongly recommends lighting from directly above, but no strong evidence indicates that this is essential. Due to problems in heating and cost of construction, the use of a glass roof, at least in northern latitudes, should be avoided until there is more evidence that such a roof is of particular benefit. In this connection it is worthy of note that large tanks that are constructed largely of opaque materials will require more light than small glass aquaria to give the same results. In direct sunlight glass aquaria become unmanageable as a result of the excess growth of algae.

Perhaps the soundest basis for calculating the amount of artificial light for an aquarium room is to determine the average amount of natural light in the room on a clear day and then adjust the intensity of the artificial light to simulate this value. Such an approach should result in the artificial light being sufficient to effectively "extend the day." Perlmutter (1962) has shown that fluorescent lighting reduces the viability of trout eggs. Incandescent bulbs should therefore be used. They may be controlled by inexpensive time clocks sold by many dealers in poultry supplies. If algal growth becomes a problem, it may be necessary to cover or to curtain portions of glass aquaria or to screen out part of the natural light and of course correspondingly reduce the intensity of artificial light.

There is more than one approach to deciding the period of illumination to be used. In their native habitat tropical fishes are exposed to 12 hours of daylight throughout the year, whereas temperate zone fishes are exposed to seasonal increase and decrease in the length of day. Tropical fishes probably fair best with a constant 12-hour day. The situation for temperate-zone fishes is a bit more complex. It is probable that for work not involving attempts to produce natural spawning, the length of day should correspond to the seasonal temperature at which the fishes are being held. When one is attempting to produce a natural spawning it may be necessary to systematically increase illumination for spring-spawners or systematically decrease it for fall-spawners. In connection with the control of spawning one should consider the possibility of using pituitary hormones. (See section on control of reproduction.)

Water Supply

TEMPERATURE: The breeding and holding of fishes requires fairly accurate control of water temperature, and experimental conditions usually prescribe a particular water temperature. Temperature characteristics of water are dependent upon the source. Well water is in the 50° to 60° F. range and remains approximately the same summer and winter. Lake, stream, and tap water are markedly affected by air temperature, although water from a spring fed stream and water from 20 or more feet of depth from a lake have a stable temperature quite similar to well water. In any case facilities should be available for heating water prior to its use. For most installations a conventional gas or oil type water heater of high capacity is the best choice. In many cases it is desirable to have duplicate water supply lines, one for cold water and the other for heated.

CARBONATE CONTENT: The minimal mineral content of water does not usually constitute a problem with the possible exception of the carbonate content.

The latter is associated with the water's ability to maintain a stable pH and especially to avoid drastic drops of pH in unflushed tanks. At a methyl orange alkalinity of less than 25 p.p.m., it is desirable to add finely ground calcium carbonate to unflushed tanks.

CARBON DIOXIDE: Carbon dioxide frequently occurs in well and spring water. Fishes commence to exhibit effects of it at 50 p.p.m., and it is advisable to maintain the concentration at a level below 30 p.p.m. even though fishes can survive at higher concentrations. The most efficient way to remove carbon dioxide is to spray the water into a louvered tower. It can also be chemically removed with hydrated lime or tris-hydro-xymethyl-aminomethane (for use of latter see Chapter 8).

IRON: Iron, principally in the form of ferrous bicarbonate, is a common contaminant of well and spring water. When this water first comes from the ground, it is clear, but upon being aerated the ferrous bicarbonate, at least in neutral and alkaline water, forms a hydrated oxide in the form of a brown precipitate. In the dissolved state, i.e., ferrous bicarbonate form, iron is a mild irritant to fishes. The hydrated oxide is not toxic but causes marked turbidity and badly stains tanks and equipment.

For low concentrations of iron (less than 5.0 p.p.m.) and for limited water requirements (less than 1,000 gallons per day) iron filters designed for home use are quite satisfactory. These filters utilize manganese zeolite which can be regenerated. For high concentrations of iron and for large quantities of water it is necessary to aerate the water and then remove the precipitated iron by employing a sand filter.

SODIUM CHLORIDE: Sodium chloride sometimes occurs in well water. There is no satisfactory means of removing sodium chloride on a large enough scale practical for aquarium use. Fortunately it is not highly toxic to most fishes, but its presence makes the water unsuited for some types of experimental work.

HYDROGEN SULFIDE: Hydrogen sulfide occurs in well water of some areas. It is highly toxic to fishes and its removal is usually not practical.

INDUSTRIAL POLLUTANTS: Industrial pollutants are of particular concern when a stream is utilized as a water supply. There are a great number of industrial pollutants many of which cannot be satisfactorily removed. It is not practical to deal with them in the present work.

CHLORINE: Chlorine is a serious pollutant of tap water. It may be removed by heating, aging, chemical treatment, or charcoal filtration. The latter method is superior to all others. Small, highly successful charcoal filters can be purchased for approximately $100.00. They may also be built since they consist of nothing more than a column of animal charcoal (approximately ⅛ inch particle size) through which water flows at a reduced rate.

Removal of chlorine by heating requires that the water be held at approximately 180° F. for at least thirty minutes. Removal by aging requires from four to five days and is at best haphazard. Sodium thiosulfate (0.05 gms. per gallon of water for 7 p.p.m. chlorine) is widely used for chlorine removal but it frequently breaks down and results in a drastic drop in pH. If sodium thiosulfate is used, overtreatment must be avoided and calcium carbonate should also be added. A number of commercial chlorine-neutralizing preparations are also available but are much more expensive than thiosulfate.

SILT TURBIDITY: Water from lakes and streams is frequently very turbid. This problem can be overcome by use of a sand filter (Figure 2). The principal requirements are a bed of sand and a back-flushing system.

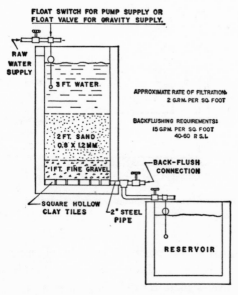

2. *Sand filter for removal of silt and other fine matter.*

BIOLOGICAL POLLUTANTS: Any organism, plant or animal, which is brought in with the water might be considered a pollutant. In some cases, these organisms can prove to be a nuisance. Well water is least apt to contain biological pollutants. Tap water is free of most biological contaminants. Lake and stream water will contain various organisms and should be filtered. The sand filter recommended for silt removal will also remove most biological pollutants. If a sand filter is not required wild fish and many other contaminants can be removed by a screen type filter (Figure 3).

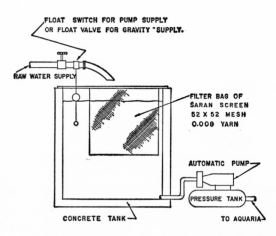

3. *Screen filter for removal of fishes and other larger biotic pollutants.*

CHOICE OF PIPE: The type of pipe (galvanized, copper, or plastic) to be used for the water supply leaves room for debate. New copper and galvanized pipe is definitely toxic to aquatic organisms, but after a period of time the insides of these pipes become coated with an inactive deposit and the "raw metal" is no longer exposed to the water. It is conceivable, however, that water of some areas would not produce a protective coating, and, further, the period required to make the pipe safe under various conditions is not known. It would thus appear that plastic is the best choice even though it may be necessary to use brass valves.

DRAINS AND OVERFLOWS: The type of tank, still-water or running-water, determines the nature of its water supply and drain. Still-water tanks that are drained or partly drained and refilled only occasionally may not require individual supply and drains whereas running-water tanks do require individual service.

A hose as a supply source and a suction type drain service are satisfactory for small, still-water tanks. The suction type drain is especially desirable since it may be used to draw off insoluble materials which accumulate on the bottom of tanks. The system consists of a steel drum equipped with a two-inch, gravity drain; a garden hose connection on one top opening and a vacuum cleaner suction hose on the other (Figure 4). In operation the vacuum cleaner creates a vacuum in the drum. The free end of the garden hose is then used to clean the bottom of the tanks and to remove as much water as desired. When the drum fills, it is emptied by the gravity drain. A pump would also do this job and for some situations might be preferred.

However, the above system has some definite advantages. It can be used for tanks containing sand, it primes rapidly and its suction force can be easily adjusted.

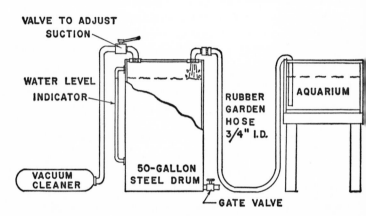

4. *A suction drain that may be used for both cleaning and draining tanks not equipped with gravity drain.*

Running-water tanks and large, still-water tanks should be equipped with a gravity drain system. The drain may be plugged by a riser pipe which serves as an overflow. It is usually convenient to combine the overflow and drain. If incoming water is colder than the tank water, it will sink to the bottom and an exchange of water will be insured. More often, however, the incoming water is the same temperature as the tank water. In this case exchange is made possible by placing a larger pipe over the overflow pipe thus forcing the water that overflows to rise from the bottom of the tank (Figure 5).

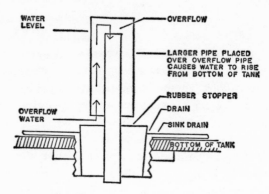

5. *Sectional view of a gravity drain and overflow for fish tanks.*

Compressed Air

An aquarium building should have a compressed air supply throughout. Each tank should have a valve and two or more should be available for larger tanks. Since this air supply is of fairly low pressure, at least three-quarter inch galvanized steel pipe equipped with needle valves is recommended. Other types of valves do not permit the degree of control of air flow that is desirable. The compressor should be a high-volume, low-pressure type. The need is for a constant air supply, hence a tank equipped with a pressure switch is not desirable. It is more satisfactory to permit the compressor to run continuously. All of these requirements are best met by an oilless, rotary pump. Most conventional piston pumps have a high pressure, low volume characteristic and are not satisfactory (see also section on Aeration). For a small number of tanks one may use any one of a number of small diaphram or piston pumps designed for aquarium use.

Tanks

FRAMELESS GLASS AQUARIA: A number of "fish-bowl" type aquaria are available in capacities up to two-gallons. These are inexpensive, durable, and a good choice for a small, nonflushing aquarium.

METAL-FRAME GLASS AQUARIA: In appearance glass aquaria are certainly the most attractive and for tanks up to 50 gallons in capacity they are generally the most satisfactory. Glass is chemically quite stable and hence does not contaminate the aquarium water. But glass tanks are expensive, approximately a dollar per gallon of capacity. With frequent draining and cleaning they develop leaks and must be disassembled and recemented. However, this problem can be greatly reduced by keeping water in the tanks even when they are not in use and by handling them only by the metal frame.

Some glass tanks have slate bottoms in which holes may easily be cut for overflows and drains. Frames may be of various metals and designs. The metal in some frames is made more rigid by fluting while some others are not fluted but are of heavier gauge metal. The latter type of aluminum or stainless steel is preferred since it is usually more easily repaired.

SOAPSTONE AQUARIA WITH GLASS FRONTS: Stone aquaria are attractive, durable and can be conveniently made much larger than metal-frame, glass tanks. Stone aquaria are expensive and suited primarily for permanent installations. The glass front of stone aquaria should be so installed as to allow for expansion and contraction, for example by framing the front of the tank with angle iron and permitting the glass to bear against this frame.

CONCRETE TANKS: Concrete tanks can be constructed in a variety of sizes and shapes but in general they are more satisfactory as large (200 gallons and up) tanks. Even for this use they do not offer the flexibility of metal tanks or the wooden tanks mentioned below. Concrete for tanks should be one part portland cement, two parts sand, and three parts gravel (¾ inch). The cement should be tamped or vibrated thoroughly to insure a tight, compact structure. After the concrete has cured, the inside of the tanks may be painted with two coats of epoxy swimming-pool paint. (This type of paint comes with the paint and catalyst in separate containers.)

If the concrete has been trowled to a slick finish, prior to painting it should be etched with a solution of one quart of muriactic acid to one gallon of water and then washed and permitted to dry. Asphalt paint is also used, but in the author's experience is not as satisfactory as the above treatment and the black finish is not always desirable.

WOODEN TANKS: Highly serviceable tanks of a variety of sizes and designs can be made of marine plywood. This type of tank can be made with one side of glass if a wooden or metal cleat is installed around one side and a metal reinforcing rod across the top of the open side. The tank, including the cleat, should be coated on the inside with fiber glass, and the glass side can then be installed with aquarium cement. The much cheaper epoxy paint recommended above for concrete may also be suitable for wooden tanks. Regardless of tank size, ¾ inch plywood is best. Joints should be carefully fitted and held together with resorcinol glue and brass nails or screws.

METAL TANKS: For a number of years the author has used galvanized stock watering tanks for holding larger fishes. They are inexpensive, available in a variety of sizes, and may be easily moved or stored. An overflow pipe may be installed (preferably in the center of the tank) to permit continuous flushing. The zinc coating of these tanks is toxic to fishes, and the tanks should be painted prior to use. As a preliminary to painting, new tanks should be treated with glacial acetic acid and then flushed. Unfortunately, the author cannot recommend a paint for metal tanks that has proven entirely satisfactory. Current tests indicate that butyl roofing paint and the epoxy paint recommended for concrete may be satisfactory.

TEMPORARY TANKS: For short experiments requiring large or unusually long tanks, it is possible to use a wooden box lined with six mil polyethylene. Since polyethylene is inexpensive and the frame may be made of scrap lumber, even large tanks can be built at nominal cost.

Selection of Experimental Fishes

THERE IS obviously a considerable morphological and physiological difference between the primitive and recent fishes, or one might say there is a phylogenetic basis for selecting an experimental fish. The bony fishes, class Osteichthyes, may be divided into the fleshy-finned fishes and the rayed-fin forms. The first group is for the most part extinct and of no interest here. The rayed-fin forms may be divided into the three groups: Chondrostei, Holostei, and Teleostei. The first group is represented in our fauna by the paddle fish (*Polyodon spatula*) and sturgeons (*Acipenser spp.*) and the second by gar pikes (*Lepisosteus spp.*) and the bowfin (*Amia calva*). Both of these groups are poorly represented in present day fauna, and forms available are too large for most experimental work.

The group Teleostei includes the great majority of our present day fishes, but even within this group there are significant phylogenetic differences. It is well to break the group into a number of subdivisions, namely: clupeids, cyprinids, cyprinodontids, and percoids. Selection of representatives from this large group constitutes a rather complex problem.

Clupeids

We know more, perhaps, about maintaining trout (*Salmo gairdner, S. trutta, Salvelinus fontinalis*) in captivity than about any other group of fishes. Various public agencies have done a great deal of research on their care and culture. Their environmental and nutritional requirements are known and rates of growth and other standards of comparison are available. Trout have some attributes which make them especially suited to experimental use. They are easily obtained from private, state, and federal hatcheries, and with plenty of fresh, cold water may be maintained in very dense populations. They feed readily upon nonliving foods including liver and heart as well as recently available pelleted food. On the negative side, trout require temperatures of about 60° F. and are best maintained in running water. To work with trout, one should have an abundance of spring or well water. Further, large trout require rather elaborate facilities.

Mud minnows (*Umbra limi* and *U. pygmaea*) inhabit small ponds and back waters, and are especially abundant in permanent swamps. They occur in the wild in the Mississippi drainage, Great Lakes drainage and Atlantic coastal areas. In nature they feed upon aquatic invertebrates. They would very likely utilize earthworms or tenebrionid larvae. As experimental fishes they have the advantage of being small, but not much is known about them as an aquarium fish. It is likely that they would reproduce in captivity.

The grass pike (*Esox americana*) occurs in streams and lakes of both the Mississippi and east coast drainages. The greater part of its diet is fishes. It is a good choice as a representative of the pike since it reaches maturity at 8 to 10 inches. It is questionable if this fish will reproduce under aquarium conditions. However, it could very likely be spawned by use of pituitary preparations.

Cyprinids

The characins are small fishes of South America, Central America and Africa. They are widely available in the United States inasmuch as they are both imported and raised as ornamental fishes. A great variety is available, but perhaps the most popular and the one about which most is written is the neon tetra, *Hyphessobrycon innesi.* One may find considerable literature on the care and culture of this group of fishes in books and periodicals dealing with "tropical" fishes. The reader is especially referred to *Exotic Aquarium Fishes* (Innes, 1938) and the *Aquarium Magazine,* The Aquarium Publishing Co., Norristown, Penn.

Characins favor a temperature of 75° to 80° F. They do quite well on a balanced ration in dry, powdered form, but an occasional meal of live food is beneficial. Characins can be fairly easily spawned in captivity. Higher temperatures (80° to 85°) and good food are conducive to spawning. As is the case with most of the "tropicals," the small size of the characins adapts them to aquarium conditions. However, the small size is a disadvantage when one wishes to weigh the fish, take a blood sample, or make injections.

The goldfish (*Carassius auratus*) has been reared as an ornamental fish for centuries, and is still raised in great numbers, both as an ornamental and as a bait species. It is widely available from commercial sources in sizes from one to five inches. The species will spawn at four to five inches in length.

The goldfish has been selectively bred for variation in finnage, color, and telescoping of eyes. Perhaps the most satisfactory stock for experimental use is one as near normal as possible in finnage, color and other morphological features. The goldfish is easy to work with. It tolerates handling, eats dried foods, and is a convenient size for blood sampling. As in the case of the trout, a great deal is known about its physiology, and standards for comparison are available for a number of its morphological and physiological characteristics.

The golden shiner (*Notemigonus chrysoleucas*) is produced commercially as a bait species, and is widely available in a selection of sizes. It feeds readily upon dry foods and is fairly well adapted to aquarium conditions. It is recommended here as suitable for short term studies such as toxicity tests.

The zebra fish (*Brachydonii rerio*), although not a native of North America, is a favorite ornamental variety and is widely available in pet shops. The zebra grows and reproduces readily in captivity. It has a short life cycle which is of particular value in some studies. It can be maintained on dried foods and is small enough, about 1.5 inches, that only the most limited facilities are required. Its shortcomings include being too small to be injected or to be weighed conveniently. Like most "tropicals" the zebra should be maintained at a temperature of 75° to 80° F.

To spawn this species, one should place several males and females in a cage of nylon marquisette suspended in an aquarium. Water depth in the tank should be only three or four inches. The cage should be suspended approximately one inch above the bottom of the tank. When the fish spawn, the eggs fall through the netting and escape being eaten. Zebras may also be spawned over gravel, but a large tank is required.

Bullheads (*Ictalurus spp.*) are available in most areas of the United States. They can be held in aquaria but are especially recommended for short term studies involving bringing experimental fishes in from the field. Bullheads will feed on meats or meals. They respond negatively to light and are probably more suitable as experimental fish when kept in subdued light. In small ponds bullheads tend to become very numerous, and by seining one may frequently obtain large numbers of uniform size.

There exists considerable interest in the culture of channel catfish (*Ictalurus punctatus*). This has resulted in the accumulation of knowledge concerning this species and has made channel catfish of all sizes available commercially. Notable among the areas where this culture is being practiced are Arkansas and Mississippi. Eggs of the channel catfish can be obtained by use of fish pituitary extract or even chorionic gonadotrophin. Catfish can be crowded and fed dried foods. It is favored by temperatures from 70° to 80° F. It spawns at a weight of two pounds or less but will attain a maximum of 20 pounds or more under natural conditions.

Small South American catfishes of the genus *Corydorus* are popular aquarium fishes. They are obtainable from dealers in tropical fishes. They can be bred in captivity but do not reproduce as readily as many other species of "tropicals."

Cyprinodontids

All of the cyprinodontids mentioned here are exotics, live-bearers, widely available, breed readily in captivity, and have been used for experimental work. They require temperatures from 75° to 80° F. and can utilize dried foods but benefit from at least some live food such as microcrustaceans.

Few fishes are more adapted to aquarium conditions than the guppy (*Laebistes reticulatus*). It reproduces rapidly, and since it is not too inclined to eat its young, no special precautions are needed to insure their survival. It has a short life cycle. Numerous colors and finnage mutations of the guppy have been preserved by hobbyists.

The platy (*Platypoecilus maculatus*) and swordtail (*Xiphophorus helleri*) are quite similar to the guppy in size, color variations and environmental requirements but do not reproduce as successfully or survive as well as guppies.

Percoids

The green sunfish (*Lepomis cyanellus*) occurs throughout most of the eastern United States. It is found in both lakes and streams. It is especially abundant in small ponds when predator fishes are not present. It is not available commercially but is easily obtained by seining.

The green sunfish will eat liver and other meats, but tenebrionid larvae or earthworms are a better choice of food for it. It does not normally breed under aquarium conditions but breeding could very likely be artificially induced. In southern latitudes, this species reaches a size of one-quarter pound or larger, but it will spawn at three or four inches. It tolerates considerable handling and does well under aquarium conditions. Larger specimens may become pugnacious, and sorting out of individuals may be necessary.

The cichlids, convenient-size experimental fishes adapted to aquarium conditions, are usually in the four- to seven-inch class. They are natives of South America and Africa, but at least some of them are widely available in the United States. Those of the genus Tilapia are of particular interest. They are raised as food fishes in Africa, and Mr. Swingle at Alabama Polytechnic has experimented with them both as food fish and as sport fish. The breeding habits of this group are rather specialized. Some members of the group carry the developing eggs and the young in their mouths.

The cichlids are quite adaptable in their feeding. They will utilize all types of meat and dried foods, and some will utilize green plants. They should be held at a temperature between 75° and 85° F. See also Gordon (1950), Nigrelli (1953), and Brown (1957).

4

Food

THERE ARE major differences between the foods which various fishes will eat. Some fishes thrive on dry meals or pelleted foods while others require fresh meat or even live food. In the following discussion we will deal with the source, use, and characteristics of representative foods.

Prepared Foods

Prepared foods are compounded from high protein, botanical meals supplemented with meat scraps, minerals, and torula yeast or distillers dry solubles. For small, surface-feeding forms such as the goldfish, golden shiner, zebra fish and guppy, the mixture is ground to a fine powder to promote floatability. For larger surface-feeding fishes, such as trout and to a lesser extent the cichlids, compounded foods are available in a pellet form which floats.

For bottom-feeders, such as catfish and bullheads,

pelleted foods that sink are available or one may make a paste of finely ground food and drop small balls of this into the tanks. Another method of feeding food in paste form is to force it through a potato ricer. This procedure greatly reduces the tendency of these foods to become dispersed in the water and cause a polluted condition.

The amount of food to be fed is of special concern in the use of compounded rations. Overfeeding results in polluted conditions that can cause the loss of fishes by oxygen depletion. Most fishes will utilize daily a quantity of food amounting to approximately three percent of their body weight. One can start feeding at this level and adjust the amount on the basis of the amount that appears to be successfully utilized. In actual practice one may measure the volume corresponding to the weight to be fed daily and then feed by volume.

DRIED FOODS: The wide availability of commercial fish feeds has eliminated the necessity of compounding one's own ration. The pelleted, commercial rations have the added advantage of the components being bound together during the pelleting process. This avoids the problem of selective loss of components by variation in solubility. The author prefers a commercial trout pellet ground to a size appropriate for the fish being fed.

GORDON'S LIVER-CEREAL RATION: Gordon (1950) suggests a mixture of a pound of beef liver, 20 tablespoonfuls of Pablum, and 2 teaspoonfuls of table salt mixed to a paste consistency. The liver is diced, a small amount of water is added and then liquidized in a food blender. The other constituents are then mixed with the liver and heated sufficiently to coagulate the liver. This mixture is refrigerated or frozen until fed.

DRIED SHRIMP: Dried, shredded shrimp is available from most tropical fish dealers. It is convenient to use but probably should be supplemented with other foods.

Live Food

In general fishes do better on live food than on any other of the various foods recommended. Live foods also cause less of a pollution problem but may be a source of parasites and diseases.

Various recommendations have been made for the culture of protozoans as food for small fingerlings, but the author favors spawning smaller fishes in large tanks and permitting the fingerlings to forage until they are large enough to feed on *Daphnia*, brine shrimp or finely ground meals. In fact most fingerlings can utilize these foods as soon as they start feeding.

Daphnia are excellent food for young fishes of all species as well as for the adults of many of the small fishes. They are also an excellent supplement to use when a dry meal is being fed. *Daphnia* can be obtained from the wild by straining fertile, stagnant water through fine mesh silk or nylon cloth. They are especially abundant in temporary pools. They occur during most of the year even under ice but appear less abundant during mid and late summer. It is practical to culture *Daphnia* in small outside pools, as for example pools of 400 to 500 gallon capacity. Their culture involves the addition of manure, hay or dried yeast to the water and then innoculating with a culture of either *D. magna* or *pulex*. *Daphnia* may also be reared indoors in large tanks. Indoors it is more convenient to use dried yeast or bone meal. The tanks should be aerated and kept at a temperature of about 70° F. One should guard against adding too much food material.

Brine shrimp (*Artemia salina*) are hatched from eggs obtained from supply houses. The eggs are available in large quantities at reasonable cost. The shrimp eggs are placed in salt water (eight tablespoons of salt per gallon of water) maintained at a temperature of 70° to 80° F. and aerated. In one or two days they hatch and are ready to be utilized. Several containers of brine shrimp hatching in sequence will produce a dependable supply of live food for small fish. This is a good food and in most cases a more dependable source of this size and type than *Daphnia*. Brine shrimp are attracted to light and may thus be separated from unhatched eggs. Microcrustaceans may be concentrated for feeding by use of a dip net or strainer made of ladies nylon hose.

Tubifex worms (*Tubifex tubifex*) occur in extremely dense populations at the water's edge of ponds and rivers which are polluted with sewage or some other concentrated organic matter. These annelids are approximately one inch in length. They are excellent food for most fishes. Tubifex can be dipped up with mud in which they stay partly buried. The mud can be washed away and the worms accumulated in pure form.

White worms (*Enchytraeus albidus*) can be cultured in trays of damp, loamy soil at a temperature of 55° to 70° F. They are fed pablum, oatmeal or bread. The tray is covered with burlap under which the worms collect and from which they can be removed for use. They tend to collect at the spots at which food is added and can be picked out in masses with tweezers. They are suitable food for all except the smallest of fishes.

Earthworms (*Lumbricus spp.*) make good fish food and may be fed whole or chopped. They can be grown in a bed of rich soil including some leaf mold and fed a variety of substances. Lard and corn meal have been recommended. The soil must be kept damp but not wet. Sand should not be included in the worm bed as the sharp edges of the grains damage the alimentary tract of the worm. The worms must be picked out of the bedding material by hand and can be used at any size.

Meal worm larvae (of beetles in family Tenebrionidae) can be cultured in chick laying mash placed in a one-inch layer over the bottom of a covered container. A 50-gallon aquarium is about the right size. A thick layer of meal tends to sour, so it should be kept thin. Larvae or adult beetles for starting the culture can be collected from the wild or obtained from a biological supply house. They will grow with little attention. Moisture may be added, if desired, on a damp cloth or cotton, but care must be taken not to wet the mash, since the larvae are removed by sifting the material through a screen. Full-grown larvae are about an inch in length, but larvae of any size may be fed. Meal worms are suitable food for larger fishes such as the green sunfish, cichlids, and bull-heads.

In many situations small fishes constitute the best food for large fishes and especially the more carnivorous forms such as the grass pike. One can use commercially available bait minnows, wild minnows, or young fishes of almost any species. The latter two groups are obtained by seining, and, if properly handled, can be held for extended periods of time and utilized as needed.

For harvesting small fishes use a ⅛ or ¼ inch minnow seine ten feet or more in length if permitted by state conservation laws. The captured fish should be penned in the net at the end of a seine haul and then, by use of a hand net, dipped into a transport facility (See Transport of Fishes). In hot weather, after the fish are in the water, a small amount of ice can be added to keep the temperature around 60° to 70° F. When the fish arrive at the laboratory, they should be acclimated to the temperature of the holding tanks (see special section below on acclimation). Frogs, tadpoles, and crayfishes are eaten by many fishes but are not necessarily any more desirable than small fishes. On a year-around basis fishes are more available.

To produce forage fishes, a pond should be freed of all fish and other competition by draining or poisoning and then stocked with adult golden shiners, fathead minnows (*Pimephales promelas*), or goldfish.

5

Diseases, Parasites, and Special Problems in Handling Fishes

FISHES SUFFER from numerous diseases and parasites. In many cases, determining the cause of a particular problem or ailment is a time-consuming job and may require special equipment and trained technicians. This is especially true of bacterial infections. Diagnosis based on gross symptoms is not satisfactory in many cases. Its reliability is related to the experience of the investigator, but, unfortunately, a great deal of experience is needed to insure even a reasonable degree of accuracy. Frequently one disease or parasite repeatedly gives trouble, and it becomes evident that considerable work in identification of the pathogen is warranted. While some treatments are effective against a number of different pathogens and thus it is possible to utilize a "shotgun" approach (e.g., formalin is effective against a number of external parasites), in some cases specific treatments are required. *Ichthyophthirius* is an external, protozoan parasite but does not respond very well to formalin. It will, however, respond to quinine.

A word of caution is in order concerning the use of various chemicals for the treatment of fish diseases. Fishes vary in their sensitivity to some of the chemicals recommended. Also some medications vary in their composition. Fishes weakened by disease are sometimes particularly susceptible to toxicants. Taking these points into consideration, it is best to treat a small lot of fish before undertaking treatment of large numbers of fishes. When fishes are susceptible to a recommended concentration, it is possible to lower the concentration and increase the time of exposure. In experimental work it is sometimes best to discard all fish, sterilize the holding facilities and start over with healthy fishes.

Many diseases and parasites of fishes have been described, but our principal interest is in the more important parasites and diseases that are apt to develop under aquarium conditions. As will become apparent, the forms that are most important are bacterial and protozoan. They are, for the most part, quite contagious and usually produce acute symptoms.

Dietary Diseases

Fishes may eat, grow, and appear quite healthy and still develop dietary diseases and die. Since fishes suffering from dietary diseases frequently exhibit low resistance to infectious diseases, the cause of death is often attributed to the latter. In a situation where fishes remain healthy for two to three weeks and then commence to show symptoms of ill health, dietary trouble should be suspected. Obviously the best

solution is a balanced ration. The greatest number of
dietary diseases occur when fishes are maintained
entirely on prepared foods rather than being fed live
foods periodically. In a separate section the preparation of balanced rations is discussed.

Mechanical Damage

Fishes are more sensitive to mechanical damage
than is commonly realized. Even slight abrasions may
result in fungal and bacterial infections. When infections become evident one or two days following handling, one should suspect mechanical damage. It is
probable that the more important infectious organisms (*Aeromonas* among the bacteria and *Saprolegnia*
among the fungi) are usually present but normally
become a problem only as a result of fishes being
mechanically damaged or becoming weakened due to
an inadequate diet.

Fright and Temperature Shock

Fishes exhibit fright as do the higher verte-
brates. They also tend to become accustomed to aquarium conditions and to being fed and handled. Following capture from the wild, fishes should be left
undisturbed for one or two days and then disturbed
as little as possible for about a week. By the end of
this time, they will have become adjusted to their new
surroundings.

Fishes are sensitive to drastic changes in temperature. When they are moved from one container of
water to another, differences in temperature should
be determined with a thermometer. If the difference
is greater than one or two degrees the fishes should

be acclimated to the new temperature. In most situations this can best be accomplished by placing fish in a plastic bag partly filled with water from their original container and then placing the bag in the tank to which the fish are to be transferred. After the temperature of the water in the bag has reached the temperature of that in the tank, the fish may be released. If the fish are too large or numerous, cooler or warmer water may be added to the old tank over a one-hour period to bring it to the temperature of the new container.

Fighting

Fishes confined in tanks are frequently killed or damaged by fighting. Fighting may occur either between species or between members of the same species. Fighting between members of the same species is much more common in some species than in others. Size difference encourages fighting, and there is little doubt that seasonal sex activity also plays a part. An aggressive fish can kill another of equal size by constantly nipping and bumping it. This is very evident in the sunfishes. A predaceous form such as the largemouth bass will often kill small fish over and beyond what it uses for food.

Viral Diseases

Viruses cause a number of diseases of fishes but we have knowledge of only two or three of these ailments. Lymphocystis is easily recognizable but does not cause heavy mortality and is of questionable importance among captive fishes. Lymphocystis infects the epidermis of fishes where it causes hyperplasia with the production of wart-like growth on the fins. Probably more important among captive fishes, especially guppies, is a dropsal condition that may be of viral origin.

There are presently no chemical treatments for viral diseases of fishes. The only approach to the problem at present is to discard the fishes and obtain new stock or utilize a different species if feasible (see Watson 1954).

Fungal Diseases

There is considerable conflict of opinion concerning the importance of fungi as fish pathogens. It is generally agreed that water molds of the genus *Saprolegnia* are the principal forms involved. It is also generally accepted that *Saprolegnia* is most important in the infection of fishes that are damaged, diseased, or suffering from malnutrition. There appears to be a relationship between temperature and fungal infections. Lower temperatures favor molds. Thus, when a mold problem becomes evident, an increase in water temperature may be a solution. Molds are also encouraged by accumulation of excess food in tanks. Molds constitute a special problem in the case of fish eggs. Successful hatching in some cases may require an increase in temperature or treatment of the eggs with a fungicide.

Various other maladies are frequently diagnosed as fungal. White discoloration on fishes is not necessarily evidence of a fungal infection. Many infections as well as certain chemical damage can cause white discoloration. The presence of the cotton-like growth of mycelia, however, indicates that *Saprolegnia* is involved.

Malachite green and copper sulphate are most commonly used as fungicides. For fishes, a copper sulphate dip of 1:2,000 for 1 to 2 minutes or a malachite green dip of 1:15,000 for 10 to 30 seconds is recommended. For the treatment of eggs, Burrows (1949) recommends an hour-long treatment of malachite green at 1:200,000 twice weekly.

Bacterial Diseases

As pointed out above, it is frequently not practical to positively identify bacterial pathogens. The matter is further complicated by the fact that all of the bacterial pathogens of fishes have not been thoroughly investigated and there exists some confusion as to the number of organisms involved. Further, little is known concerning the susceptibility of different species of fishes to the different pathogens. Despite these problems it is still of value to outline the commonly recognized bacterial diseases. The reader is also referred to the following important review papers: Griffin (1954), Snieszko (1954), and Rucker *et al.* (1953).

Furunculosis, caused by the bacterium *Aeromonas salmonicida,* has been rather thoroughly studied. The salmonid fishes are especially susceptible to it, although it is also suspected of infecting warm water fishes. The bacterium is a short, nonmotile, gram negative rod that shows bipolar staining. Host symptoms include bursting of capillaries, resulting in red spots, reddening at the base of the fins and a bloody fluid from the gut. Septicemia is usual, the kidneys are swollen, and furuncules may or may not develop. Sulfamerizine at the rate of 8–10 grams per 100 lbs. of feed is recommended as chemical treatment. *A. sal-*

monicida survives only a short time outside of the fish. Carrier fishes are important in its spread (see Snieszko 1958a).

The species *A. liquefaciens* and related forms appear to cause several pathogenic conditions in fishes. The different forms appear very similar both morphologically and culturally but produce diseases that are quite different. All of the forms are motile, gram negative rods that show bipolar staining. They are credited (by some investigators) with causing infectious dropsy, which involves the accumulation of serous fluids in the tissues, and a fatal surface infection of fish following handling. The surface infection is readily controlled by 20 p.p.m. of water-soluble terramycin (veterinarian grade) for 24 hours or more. For dropsy one should use chloromycetin or terramycin in the food or by injection. When used in the food, the antibiotics are fed at the rate of 1.0 milligram per 15 grams of fish for a period of four days. When injected, the dose is 0.025 milligrams per gram of fish. Control of bacterial infections by either injection or contact treatment with terramycin is frequently very successful.

At least certain forms of *A. liquefaciens* are free-living and constitute a normal part of the bacterial flora of natural waters. The development of the surface infection of fishes appears to be associated with minor damage to the fish during harvest.

Columnaris, an important disease of both warm water and cold water fishes, is caused by the bacterium *Chondrococcus columnaris*. It produces grayish-white areas on a fish's head, gill, fins, body and frequently in the mouth. These spots change to shallow ulcerations, and the fins may become frayed. The infection eventually becomes general and the bacterium can be found infecting the internal organs. Microscopic diagnosis is possible since the bacterium has a characteristic appearance: a long, thin rod with an oscillating motion and a tendency to congregate into columns (see also Snieszko 1958b).

For the control of columnaris, Davis (1956) recommends 1:2,000 copper sulfate dip repeated two or three times at 12 to 24 hour intervals when the disease is in its early stages. Fishes having extensive lesions should be discarded. O'Donnell (1941) recommends a treatment of 1:15,000 solution of malachite green for 10 to 30 seconds.

Ulcer disease caused by the bacterium *Hemophilus piscium* is not dealt with in detail here because it appears to be fairly well limited to trout. Persons interested in this disease are referred to Piper (1958).

Protozoan Parasites

The following is not meant to be a complete treatment of the protozoan parasites of fishes. The more common forms are listed to serve as an example of the symptoms, identification and treatment of this type infection. The reader is referred to the excellent review by Davis (1956). Much of the information given here is based on Mr. Davis' book.

Costia nectris and *C. pyriformis* are external parasites affecting both coldwater and warmwater fishes. They cause a bluish or gray film to form on the fish's body, but diagnosis should be made by microscopic examination. Costiasis is considered to be a rapidly fatal disease but has been successfully controlled by

three different procedures: a one-minute dip in a
1:500 solution of acetic acid, 12-hour treatment of 1 to
20,000 formalin, and a one-hour treatment in a 1:500,-
000 solution of pyridylmercuric acetate (PMA).
Some fishes are sensitive to these treatments (see also
Fish 1941).

The genus *Chilodon* includes a number of fish parasites. As is the case with many parasites, they become
a serious problem when fishes are crowded or become
weakened. The most evident symptom is a tendency
for the host to become weak and emaciated. Diagnosis is by microscopic examination of scrapings from
the infected fish. The organisms are usually concentrated on the gills. Control is by a short dip in a 3
percent solution of common salt, a one-minute dip in
glacial acetic acid diluted to 1:500, or a 12-hour treatment of 1 to 20,000 formalin.

Ichthyophthirius multifiliis is one of the easiest of
diseases to diagnose. It appears as uniformly-shaped
white specks scattered over the host. Both the adult
form and the encysted form are visible without mag-
nification, but identification should be verified by
microscopic examination of the fins and body of the
fish. The control of "ich" is complicated by the occurrence of an encysted stage which is not affected by the
usual treatments for external parasites. The parasite
may be partially controlled by placing the fish in running water for three or four days. (Maintaining the
aquarium water at 85° F. for several days will control "ich" on tropical fishes.) But where fish are confined in smaller tanks, a more convenient treatment is
with quinine sulfate. Five-tenths grains or 0.032 grams
per gallon of water is added to the holding tank, and

the water is left unchanged for three weeks. Quinine is toxic to plants; hence, any plants present must be removed.

Oodinium limneticum is a dinoflagellate which causes a sickness commonly referred to as gold dust disease. Infected fish have the appearance of being covered with a golden dust. *Oodinium* has both attached and free-swimming stages. The attached stage is parasitic. It covers the fins and general body surface of the fish. Infected fish become emaciated, and at least younger fish suffer heavy mortality (see Jacobs 1946). Diagnosis is by the gold-colored growth (*Oodinium* possess a yellowish pigment) on the fish's body and by the identification of the attached form.

Oodinium is especially important on young fish and more delicate species. In some cases this parasite lives on adult fish without causing great damage, but it drastically affects young fish of the same species. The author has successfully used a malachite green dip at 15 p.p.m. for 2 minutes for control of the encysted form of *Oodinium*.

Although protozoans of the genus *Trichodina* may heavily infest fishes, they normally do not seem to cause pathological symptoms. Their occurrence is easily detected by microscopic examination of the fins of the fish. They can be removed by a 12-hour treatment with 1 to 30,000 formalin.

Parasitic Worms

Most of the parasitic worms of fishes have complicated life histories involving two or three hosts. Aquarium conditions are such as to prevent these parasites from being a problem. Trematodes of the genus *Gyrodactylus* and *Dactylogyrus,* however, constitute an exception. Parasites of both these genera complete their life cycle on a single fish host and are capable of building up heavy infestations. Transmission is by contact. *Gyrodactylus* occurs mostly on the fins and body of the host fish. It gives birth to well-developed young. *Dactylogyrus* parasitizes the gills of fishes. It is an egg-layer and reproduces somewhat more slowly than *Gyrodactylus.* Control is by a 12-hour exposure to 1 to 20,000 formalin. A two-minute dip in 1:400 solution of acetic acid has also been recommended.

Parasitic Copepods

Two important warm-water parasitic copepods may be encountered on captive fishes. They belong to the genera *Argulus* and *Lernaea.* The fish louse, *Argulus spp.,* moves at random over the host's body. It may be more than a quarter of an inch in diameter and, of course, visible to the unaided eye.

The anchor worm, *Lernaea spp.,* once established, is a considerably more serious parasite than the fish louse. It has a complex life cycle involving a free-swimming stage, reproductive stage on the gills of fishes, and the female has an attached stage during which the eggs are developed. The latter stage is usually observed. This form may be as much as a half inch in length. It appears as a boney splint protruding from the body of the host. It does considerable mechanical damage to certain fishes such as the goldfish and golden shiner.

No entirely satisfactory control exists for the parasitic copepods. Their tough body covering protects them from the usual treatments for external parasites. The free-swimming stage of the anchor worm is sensitive to strong salt solutions or long exposure to a formalin solution. The insecticide benzene hexachloride at a concentration of 0.1 p.p.m. of the gamma-isomere used in the water in which the fish are contained has been recommended. It is quite toxic to fishes and cannot be highly recommended for use on confined fishes. Relevant to diseases and parasites see also Van Duijn (1956) and Nigrelli (1943). Relevant to physiology of fishes and habitat requirements see Brown (1957).

6

Aquarium Plants and Other Miscellaneous Considerations

A PLANT suitable for use in aquaria must be small and in most cases tolerant of reduced light. Numerous plants are used by hobbiests since they are especially interested in the aesthetic effect produced by variety. The author has selected the more widely available plants and representatives of different growth habit. The four plants mentioned are available from most "tropical" fish dealers.

Water sprite (*Ceratopteris thalictrodides*) is one of the most versatile aquarium plants available. It will grow equally well in a substratum and totally submerged or without substratum and floating at the surface. It reproduces rapidly by asexual means and grows rapidly.

The vallisnerias (*Vallisneria americana* and *spiralis*) are grass-like plants that require a substratum of sand or gravel. Their asexual reproduction is rapid if sufficient light is present. *V. americana* grows more rapidly and makes a larger plant than *V. spiralis.*

The Amazon sword plant (*Echinodorus rangeria*) is a broad-leafed plant that requires a substratum. It reproduces slowly, but each plant reaches a size of a foot or more in diameter. In general, it is a more expensive plant. Under good conditions it puts out runners which can be rooted with ease.

Control of Reproduction

When held at the proper temperature and adequately fed, many of the small fishes will reproduce in aquaria without any special inducement. Some fishes will satisfactorily develop gametes but fail to actually spawn. Still other species will not develop eggs under aquarium conditions. Some fishes can be stripped, while others can be stripped after a preliminary injection of pituitary. In addition fighting between sexes and egg and fry eating add to the complexity of the problem of getting fishes to successfully reproduce in captivity. Some of the techniques for dealing with these problems are discussed below.

The development of gametes and spawning are under the control of hormonal products produced by the pituitary gland. The pituitary in turn is stimulated by light and temperature. In addition, proper nutrition and freedom from crowding (Swingle 1956) are essential for the development of the gametes. If gamete development and spawning cannot be obtained by adjustments of food, light, temperature and population density, it is possible, at least in numerous species of fishes, to induce both by injections of pituitary products. To stimulate gamete production a series of injections of ground fish pituitary or A.P.L. is used. Spawning by "ripe" fishes can usually be triggered by one injection or in some cases simply by moving the fish to fresh water.

Fish pituitary is prepared by obtaining pituitaries from large fishes (carp has been widely used) from the wild, preferably in the spring just prior to spawning time. The excised pituitaries are dehydrated in acetone, dried and stored in tightly sealed containers. They will remain active for one or two years. For use the dried glands are ground, suspended in sterile water and injected in the coelom. If inflammation becomes a problem, penicillin G may be mixed with the pituitary suspensions. Two problems are associated with the use of fish pituitary. It is not generally available and must be obtained from the wild, and the dosage is difficult to calculate inasmuch as the level of activity of the material is usually not known. About the only approach to determining the dose for individual conditions is by trial and error.

Recently Sneed and Clemens (1959) reported chorionic gonadotrophin to be satisfactory for triggering spawning in a number of fishes. For the channel catfish they recommend 800IU per pound of fish.

The use of pituitary is greatly simplified if ripe fish are taken from the wild or from outside pools and the pituitary relied upon solely to trigger spawning. For an excellent review of the use of pituitary in spawning fishes the reader is referred to Atz and Pickford (1959).

EATING OF EGGS AND FRY: Some fish such as the swordtails and platies will reproduce readily but are inclined to eat their young unless vegetative cover is available in the tank or unless the adults are kept in nets which permit the young fish to escape. Zebra fish and goldfish will lay eggs but frequently eat the eggs immediately after they are laid. To avoid this problem, gravel may be used in zebra tanks. The eggs fall into the interstices of the gravel and are protected. Goldfish may be permitted to spawn on mats of dried spanish moss or grass roots, and immediately after spawning is complete the mats are moved to another tank.

FIGHTING BETWEEN SEXES: Fighting is often associated with reproduction. In some cases a change in size relationship between the sexes will reduce this problem. At times permitting the female to reach a more advanced state of development before placing the male with it will correct the problem. Mr. Shell of Alabama Polytechnical Institute in breeding *Tilapia* uses small females and large males. They are placed in a trough with a divider which permits the smaller females to pass but not the males. The female thus associates with the male at will (Shell unpublished).

STRIPPING: Stripping is a routine procedure in the culture of trout and salmon and is practical for some other, but not all, fishes. It has the advantage of giving exact information as to the time of fertilization and age of embryo. Crosses can sometimes be made by this method between forms which otherwise will not mate. Using the stripping technique, fishes which often eat their own eggs may be used for experimentation. (Strawn, Kirk, and Hubbs 1956).

Ripe males and females are used for stripping. The process involves gently pressing the mature eggs from the female into a suitable container, adding sperm similarly pressed from a mature male and gently brushing the milt over the eggs with a feather or the anal fin of the male. A short time following fertilization, the eggs should be spread out and excess milt and mucous carefully washed away. The eggs must then be incubated at a suitable temperature and aeration until they hatch. Dark or cloudy eggs are dead and should be removed since their presence may encourage the growth of fungus.

Anesthetization of Fishes

It is sometimes desirable to anesthetize fishes prior to working with them, or it may be desirable to anesthetize them to decrease the difficulty of capturing them from aquaria. Carbon dioxide either as compressed gas, dry ice, or from the addition of sodium bicarbonate followed by the addition of sulfuric acid diluted 1 to 500 (for further details on the bicarbonate method see Fish 1942) is an effective anesthetic especially suited to permit capture of fishes in aquaria. Fishes are anesthetized at concentrations above 150 p.p.m. Anesthesia occurs in a matter of minutes; fishes satisfactorily recover at lower concentration but will die if permitted to remain in high concentrations of carbon dioxide. An anesthetizing concentration of carbon dioxide can be removed from water by one or two hours of aeration.

Quinaldine, a coal tar derivative, is an inexpensive and effective fish anesthetic. At seven p.p.m. it will quickly incapacitate fishes and keep them incapacitated for extended periods of time. Upon being removed from the quinaldine solution, the fishes recover rapidly. Quinaldine is not known to be highly toxic to warm blooded animals and is not considered to be carcinogenic (Muench 1958).

M.S. 222-*Sandoz* has been widely used as an anesthetic for fishes. It has a number of desirable traits but is considerably more expensive than quinaldine. It is used at concentrations as high as 1 to 1,000. For more details the reader is referred to the bulletin "M.S. 222-Sandoz, The Anesthetic of Choice in Work with Cold-Blooded Animals," published by Sandoz Pharmaceuticals, Hanover, N. J.

Urethane is an effective drugging agent for fishes and has been widely used, but it is not recommended due to its pronounced carcinogenic property.

Sodium cyanide will satisfactorily drug fishes at 0.5 to 1.0 p.p.m. When removed to fresh water, fishes drugged with cyanide recover rapidly and exhibit no after effects (Bridges 1958). Sodium cyanide is inexpensive, but, of course, is very dangerous. Death can result from swallowing small amounts of the salt and from breathing low concentrations of the gas, hydrogen cyanide. Hydrogen cyanide is produced in quantity when the salt is placed in an acid solution.

Sodium amytal has been used as a fish anesthetic, but its action is so slow as to make it unsatisfactory for most uses.

Sterilization of Tanks and Equipment

Chlorine is a highly effective and practical sterilizing agent for aquarium room use. It should be used at 200 p.p.m. for 30 minutes. Chlorine is available as compressed gas, calcium hypochlorite, and dissolved gas. Hypochlorite releases 30 to 40 percent of its weight in gaseous chlorine. The most available form of dissolved chlorine is common household bleach. The choice of form depends upon the size tanks to be sterilized and the amount of sterilizing to be done. The gaseous form is the cheapest but is less available, harder to handle, and dangerous. Calcium hypochlorite is fairly cheap and convenient to use. Chlorox, Hilex, and other household bleaches are convenient to use but are more expensive.

Chlorine solutions may be used in two ways. For floors, benches, large tanks, etc., the surface may be rinsed or mopped several times with a 200 p.p.m. or stronger solution of chlorine. Small tanks may be filled with the chlorine solution, and small objects may be submerged for thirty minutes. In some cases it is well to maintain a crock of chlorine and a crock of neutralizing solution (discussed below) to permit dipping of nets and other equipment which might transfer disease from tank to tank.

Chlorine is effectively neutralized by sodium thiosulfate (photographic "hypo"). Following sterilization of tanks and other equipment with chlorine, they should be rinsed with a 100 p.p.m. solution of this compound. The tanks should then be rinsed with water after which they may be used.

In cases where the use of chlorine is not desirable, Roccal, a germicidal agent sold by Winthrop-Sterns, may be used.

Circulation and Aeration of Water

Oxygen is only slightly soluble in water; even at saturation the available reserve is limited, and oxygen must be continually replenished. In addition the removal of gaseous waste such as carbon dioxide and ammonia is necessary. In nature both replenishment of oxygen and removal of gaseous waste is by exchange at the water's surface and by photosynthesis. The former is greatly aided by surface agitation by wind. In the aquarium environment fishes are usually more crowded and surface exchange and photosynthesis do not always proceed at a normal rate. It is thus customary to aerate tanks by compressed air, mechanical agitation, or by circulation. Air pumps or compressors vary greatly in their capacities and design. It has been pointed out earlier that air pumps for aquaria aeration should be of a high volume, low pressure type. Further, the need is for a continuous flow of air. A number of small pumps on the market are made specifically for aerating aquaria. Some of these units are a diaphram type while others are piston type. They can adequately aerate several tanks, however, for numerous tanks and for larger tanks it is advisable to obtain a rotary pump of approximately one-quarter horsepower.

Closely associated with the use of compressed air for aeration is the use of air dispersement devices. Air stones are the conventional air-dispersing device, but the author has made excellent dispersers from thick-wall, one-quarter inch diameter, polyethylene tubing perforated with a fine needle.

For tanks in the 300-gallon class and up, aeration can more effectively be accomplished by the use of agitators (Figure 6). Suitable agitators are available from companies dealing in bait minnow supplies.

Running water may be sprayed or splashed to aerate it. If circulating or flushing is being done, a good deal of aeration can be accomplished by this means.

The air-lift type water pump is widely used by aquarium hobbiests. It is simple and inexpensive, and

6. *Agitator for aerating aquaria and hauling tanks.*

each tank may be equipped with a separate pump. Its principal application is in low pressure filtration of small tanks. The principle of the air-lift pump is illustrated in Figure 7.

Another pump of considerable use in aquarium work is a small, submersible type centrifugal. These are available in capacities as low as 82 gallons per hour and pressures equal to 7 feet of head. For draining unusually large tanks a household type, submersible, sump pump is very effective. The next best choice is a self-priming centrifugal. In most cases piston and centrifugal pumps which are not self-priming are not satisfactory.

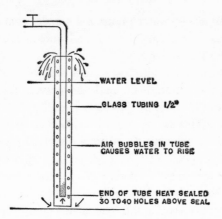

WATER LEVEL

GLASS TUBING 1/2"

AIR BUBBLES IN TUBE
CAUSES WATER TO RISE

END OF TUBE HEAT SEALED
30 TO 40 HOLES ABOVE SEAL

7. *The air-lift pump is of value in filtering water of small individual tanks.*

Filtering Aquarium Water

Prior to actual use in aquaria, water may be filtered to remove residual chlorine, present in most city water, or may be filtered to remove suspended materials as well as some plants and animals when the water supply is from a lake or stream. The types of filters used for these purposes are discussed in Chapter 2. The present discussion deals with filtering of aquarium water for reuse.

It should be noted that in most cases flushing is more desirable than filtering and reuse. But there are, of course, situations which make filtering desirable. Filtering of aquarium water removes suspended material, clarifies the water and may remove dissolved metabolic waste such as ammonia and carbon dioxide if a proper filtrant is used. The most commonly used filtering system consists of an individual filter for each tank. These small filters are operated in conjunction with an air-lift water pump and usually contain glass wool as a mechanical filter and animal charcoal to absorb dissolved gases.

A more recent filtering system removes water from below a substratum of sand and gravel. This causes the water to circulate downward and through the substratum. This is, of course, a mechanical filter and cannot be expected to remove dissolved gases.

The first type of filter also has definite limitations. In the conventional design the pressure moving the water through the filter is furnished by the column of water above the filtrant. The air-lift pump installed in an aquarium can only lift water three or four inches. Thus, the column of filtrant must be short and the filtrant must be coarse. When one is dealing with tanks of a hundred or more gallons a small submersible pump may be used to lift the water several feet and provide pressure for filtering through longer columns and finer filtrants.

It is, of course, also possible to construct a pressure filter which would involve the use of a closed filter with water circulated through it by a submersible type centrifugal pump.

Synthetic Water

The mineral content of water varies drastically both in regard to kinds and amounts. Some minerals have a pronounced effect upon the physiology of fishes and upon materials added to the water for experimental purposes. The problem of variation in mineral content has resulted in conflicting reports concerning the effect of toxicants on fishes and on parasitic organisms. To avoid this variable there appears to be a strong argument for utilizing a standard synthetic water for many of the tests done now in whatever water happens to be available.

In many cases the investigator is primarily interested in the effect of toxicants or other factors under local water conditions. If such is the case he will use local water even though the resulting observations may not be comparable to those based on work in other waters. In such a situation it would be desirable to run duplicate tests utilizing a standard or reconstituted water for one and local water for the other.

7

Chemical Variables of Aquarium Water
and Their Determination

IN THE present work we will deal only with those chemical variables which are most apt to affect the maintenance of fishes in aquaria and which can be readily measured. We have omitted variables such as acidity, salt content, and various widely occurring compounds such as the sulfates. For a consideration of these variables the reader is referred to the following works: *Standard Methods for the Examination of Water, Sewage, and Industrial Waste,* Tenth Edition, American Public Health Association, *et al.,* 1955; and *Limnological Methods,* Welch, 1948.

Dissolved Oxygen Determination

Oxygen is not very soluble in water, yet fishes are dependent upon the small quantities that are present. Dissolved oxygen in water is replenished by diffusion from the atmosphere and photosynthesis of submerged, green plants, especially algae. Dissolved oxygen is depleted by respiration by plants and animals. If some pollution is present, such as an accumulation of uneaten food, bacteria may rapidly deplete the available oxygen present in water.

Water's oxygen-holding capacity varies inversely with the temperature; i.e., the higher the temperature the less the amount of oxygen that the water will hold. The following relationships illustrate this: At temperatures of 50°, 60°, 70° and 80° F. the oxygen required for saturation is 11, 10, 9, and 8 p.p.m. respectively.

An oxygen concentration of 3 to 5 p.p.m. is adequate for warmwater fishes and many can survive at lower concentrations.

TAKING WATER SAMPLES: Water samples used for oxygen determination must not be exposed to air during sampling. To obtain a nonaerated sample from an aquarium the simplest procedure is to use a small rubber tube to siphon water into a 250 ml., glass-stopper type bottle. One end of the siphon tube should be placed in the aquarium and the other end near the bottom of the bottle and the water permitted to overflow three times the volume of the sampling bottle. The sample should be analyzed immediately after it is taken.

REAGENTS REQUIRED: (1) Manganous sulfate solution (dissolve 480 g. $MnSO_4 \cdot 4H_2O$ in 100 ml. of distilled water and dilute to 1 L.). (2) Alkaline-iodide sodium azide reagent (add 700 g. KOH, 150 g. KI to 950 ml. of distilled water; permit to cool; Add 10 g.

NaN_3 to 40 ml. of water; slowly mix the two solutions). (3) Starch indicator (2 g. soluble starch dissolved in 350 ml. hot water). (4) Sulfuric acid (concentrated H_2SO_4, sp. gr. 1.84). (5) Sodium thiosulfate
solution (dissolve 6.205 g. of $Na_2S_2O_3 \cdot 5H_2O$ in 500
ml. of freshly-boiled distilled water; dilute to 1 L.).

Sodium thiosulfate solution is unstable. Its stability
is greatly improved by storing in a dark bottle, re-
frigerating when not in use, and by adding 5.0 ml. of
chloroform per liter. Not only must sodium thiosulfate
solution be standardized following preparation, but it
must also be standardized prior to use depending
upon length of time in storage. If it is refrigerated,
standardizing at two-week intervals is adequate. If it
is not refrigerated, it should be standardized weekly.
Standardization is accomplished by the procedure
given below.

Oven dry a small quantity of reagent grade potassium dichromate ($K_2Cr_2O_7 \cdot 4H_2O$) crystals in oven at
130 degrees C. for 30 minutes, cool in a dessicator
and weigh out 1.226 g. Dissolve and dilute to 1 L. in a
volumetric flask.

To 10.0 ml. of the above potassium dichromate
solution slowly add 1.0 ml. of alkaline-iodide sodium
azide reagent then add a drop at a time 1.0 ml. of concentrated sulfuric acid. The acid should be added
slowly enough to avoid any odor of free iodine and
heating. Using the thiosulfate to be standardized,
titrate this solution as an oxygen sample (see procedure below) to a pale straw color, add starch and
titrate until the blue color fades completely. The number of milliliters of sodium thiosulfate solution divided
into 10 will give a correction factor by which thiosul-

fate readings should be multiplied to give dissolved oxygen in p.p.m.

TITRATION PROCEDURE: In the following procedure a separate pipette is used for each reagent, and each reagent is added below the surface.

To the 250 ml.-water sample add 1.0 ml. of the manganous sulfate solution. Immediately add 1.0 ml. of the alkaline potassium iodide sodium azide solution, stopper and invert twice to mix. Permit the resulting precipitate to settle. A white precipitate indicates very low oxygen, and a brown precipitate indicates the presence of oxygen. After the precipitate has settled (one or two minutes) add 2.0 ml. of concentrated sulfuric acid by letting the acid run down the neck of the bottle into the sample. Again stopper and invert to mix.

Measure out 200 ml. of the prepared sample and pour this quanity into a clear flask or beaker. By use of a burrette titrate the thiosulfate into the 200-ml. sample until the brown color becomes a pale straw color. Then add starch indicator which will turn the sample blue. The end point of the titration is reached when the blue color fades completely and does not return for a period of 30 seconds. The number of milliliters of sodium thiosulfate used times the correction factor from standardization is equal to the p.p.m. of dissolved oxygen in the sample.

Free Carbon Dioxide Determination

Carbon dioxide in aquarium water is produced by all organisms present, but under different conditions different organisms are more abundant and hence more important. Thus, when the amount of illumination is excessive, algal growth may become pronounced. During hours of illumination all carbon dioxide present may be utilized by algal photosynthesis, but during periods of darkness photosynthesis ceases, respiration continues and carbon dioxide concentrations go up. Another common source of excess carbon dioxide is from organic pollution such as results from over-feeding. In this case high carbon dioxide is due to bacterial respiration.

In most cases carbon dioxide does not constitute a problem in holding fishes in aquaria. At a concentration of 150 p.p.m. it will anaesthetize fishes and at higher concentrations it is usually fatal, but these high concentrations do not commonly occur. Carbon dioxide is quite soluble, but low values (less than 20 p.p.m.) are more characteristic of aquarium water.

TAKING WATER SAMPLE: Siphon 100-ml. water sample into a graduate. With a minimum of agitation transfer this sample to a beaker.

REAGENTS REQUIRED: (1) Sodium hydroxide solution N/44 (dissolve 0.909 g. of NaOH in 500 ml. of distilled water and dilute to 1 L.; standardize against any standardized acid solution). (2) Phenolphthalein indicator (0.5% solution in 50% alcohol).

TITRATION PROCEDURE: To a 100-ml. sample of water add 10 drops phenolphthalein indicator and then titrate with the sodium hydroxide solution to a faint, permanent pink. The milliliters of sodium hydroxide solution used multiplied by 10 and by the correction factor from standardizing gives the p.p.m. of free carbon dioxide.

Chlorine Determination

Chlorine is almost always added to municipal water supplies. Tap water may contain a residual concentration of chlorine as high as 7.0 p.p.m. (1 to 2 p.p.m. is fatal to fishes). Since chlorine is so highly toxic to fishes, the existence of any chlorine in aquarium water is undesirable. On this basis a qualitative test for chlorine seems as satisfactory as a quantitative one. Thus the simple test given here is qualitative.

TAKING WATER SAMPLE: Water samples for chlorine determination need not be handled with special care. They may be dipped from the tank or drawn from the faucet into a graduate. One should bear in mind, however, that chlorine content of tap water varies throughout the day, with the season, with the length of time the water stands in the plumbing system, and with variations in the treatment policies at the water plant.

REAGENTS REQUIRED: (1) Acid solution (dilute 500 ml. of glacial acetic acid to 1 L). (2) Potassium iodide solution (dissolve 75 g. of reagent grade, iodate and chloride-free KI and dilute to 1 L). (3) Starch indicator (2 g. soluble starch in 350 ml. hot water).

TESTING PROCEDURE: Using a pipette, put 10 ml. of acetic acid solution in a flask, add the same quantity of potassium iodide solution, then pour in

200 ml. of sample and mix. Final pH of sample should be between 3.0 and 4.0. If a brown color results the chlorine content of the sample is quite high. If a brown color does not occur, add 1 ml. of starch solution. If a blue color is produced, a small but still dangerous amount of chlorine is present. If no blue color develops the sample is free of chlorine.

ALTERNATE METHOD OF TESTING FOR CHLORINE: The indicator orthotolidine is also widely used for determining the presence of free chlorine. Use a 10 ml. sample and add 1 ml. of indicator. The production of a yellow color demonstrates chlorine to be present.

Hydrogen Ion Activity

Fishes tolerate a pH range of 5.0 to 9.0 with no indications of difficulty. At greater extremes the mucus covering of their bodies commences to coagulate and mortality occurs. Most water has enough buffering ability to prevent the development of pH values outside the safety range, but occasionally extremes can occur and in some experimental work, such as testing the toxicity of a chemical, careful pH control is required. There are a number of satisfactory colorimetric kits available which may be used for pH determination in connection with aquarium work.

Alkalinity

In aquarium work a measure of alkalinity is essentially a measure of the carbonates. The carbonates are of interest because of their ability to stabilize pH and because variations in the carbonate content can considerably affect experimental conditions. The determination of alkalinity sometimes involves two titrations rather than one. If the pH of the water is above 8.3, phenolphthalein alkalinity is said to be present; whereas, the alkalinity occurring between 4.5 and 8.3 is called methyl orange alkalinity.

Phenolphthalein Alkalinity

REAGENTS REQUIRED: (1) Sulfuric acid solution .02N (dissolve 0.6 ml. concentrated H_2SO_4 [sp. g. 1.83] in 200 ml. of distilled water and make up to 1 L.; To standardize weigh out 1.060 g. of reagent grade anhydrous sodium carbonate [Na_2CO_3], dissolve in 400 ml. of distilled water and make up to 500 ml.). The .02N sulfuric acid is titrated against an equal volume ume (10 or 20 ml.) of the carbonate solution using phenolphthalein as an indicator. The correction factor for the .02N sulfuric acid is calculated by

$$\frac{\text{ml. of } Na_2CO_3}{\text{ml. } H_2SO_4 \text{ used in titration.}}$$ (2) Phenolphthalein indicator (0.5% solution in 50% alcohol).

TITRATION PROCEDURE: Add a few drops of phenolphthalein indicator to 100 ml. of sample. If a pink color appears, titrate with .02N sulfuric acid to an end point where the pink color disappears. The milliliters of acid multiplied by 10 and the correction factor equals p.p.m. of phenolphthalein alkalinity expressed as p.p.m. of calcium carbonate.

Methyl Orange Alkalinity

REAGENTS REQUIRED: (1) Sulfuric acid solution .02N (same as for phenolphthalein alkalinity).

(2) Methyl orange indicator (0.05% aqueous solution).

TITRATION PROCEDURE: The same sample used to determine the phenolphthalein alkalinity can be used to determine the methyl orange alkalinity if the phenolphthalein end-point has been reached accurately or if no pink appeared when the phenolphthalein indicator was added. Add several drops of methyl orange indicator, then titrate with the .02N sulfuric acid to the first definite change from yellow to salmon pink or pinkish orange color. The number of milliliters of sulfuric acid solution used multiplied by 10 and the correction factor is the p.p.m. of methyl orange alkalinity expressed as p.p.m. of calcium carbonate.

Ammonia

Ammonia is perhaps the best index of the quantity of accumulated metabolic waste in aquaria, and studies involving an interest in metabolic waste would likely require an analysis of ammonia. However, due to the difficulty of making ammonia determination it is felt best to omit the procedure here and refer the interested reader to *Standard Methods for the Examination of Water, Sewage, and Industrial Waste,* Tenth Edition, American Public Health Association, *et al.,* 1955.

8

Transport of Fishes

THE TRANSPORT of fishes involves a complex of considerations. When fishes are excessively stimulated as in the case of handling and transport, their metabolic rate rises to its probable maximum which is four to five times as great as their minimal metabolic rate. The increase in metabolic rate creates a greater oxygen demand and increases the output of metabolic waste. During a period of excitement fishes develop an oxygen debt that may require several hours to repay. Excited fishes may also mechanically damage themselves. For reasons of weight economy it is necessary to greatly crowd fishes during transport. Crowding results in a rapid depletion of dissolved oxygen and an increase in carbon dioxide and ammonia. In addition, the water becomes fouled from regurgitated food, feces, and mucus.

Temperature

Although some fishes are sensitive to low temperatures, many can be handled and transported much easier at temperatures of 55° to 65° F. than at higher temperatures. The metabolic rate is lowered and loss of scales reduced in some species. Cooler water has a higher oxygen-holding capacity and the rate of reproduction of putrefying bacteria is reduced.

In warm weather lower temperatures are most easily maintained by the use of ice. It is, of course, of value to utilize an insulated tank. If the temperature at which fishes are to be transported is different from the temperature at which they are being held, they must be slowly and carefully acclimated to the new temperature. Failure to acclimate fishes to a new temperature may result not only in acute symptoms but also in delayed symptoms.

Oxygen

When fishes are crowded, a shortage of oxygen is most apt to cause early mortality, and even the crudest transport facilities include means of aeration. One may use compressed air, compressed oxygen, baffles that agitate the water as a result of movement of the transport unit, circulation of the water, and mechanical agitation.

Compressed air, in addition to oxygenating the water, probably washes out a considerable quantity of ammonia and carbon dioxide. But a large compressor is needed to maintain a high oxygen level, and in some cases warming of the water by the continuous flow of warm air is a problem. Compressed oxygen is in some respects convenient but expensive. It is also questionable if it has the washing ability of compressed air since a lesser flow is commonly used. It is a good stand-by method and for some transport facilities may constitute a desirable supplement to other aerating devices.

Agitation by baffles dependent upon the movement of the transport unit is obviously of supplementary value. Aeration by circulation, which usually involves withdrawing the water from the bottom of the tank and spraying it back on the surface, also eliminates considerable carbon dioxide and ammonia and makes possible filtration. Mechanical agitators widely used in the transport of bait minnows are available in models that operate on 110-volt domestic current or 12-volt automotive current. As a simple, fool-proof piece of equipment that can hardly be improved on, the author considers them the most satisfactory aerating device for all size tanks.

Carbon Dioxide

When fishes are crowded, carbon dioxide can accumulate rather rapidly. Higher concentrations of carbon dioxide interfere with the fishes' utilization of oxygen and cause a decrease in pH. Fish exhibit signs of anesthesia at 40 to 50 p.p.m. carbon dioxide, and McFarland and Norris (1958) demonstrated delayed mortality that appeared to be associated with concentrations even below 20 p.p.m. McFarland and Norris also investigated the control of decrease in pH during transport by the addition of tris-hydroxy-methyl-aminomethane referred to as "tris-buffer." They considered that the decrease in pH during transport is primarily a result of an increase in carbon dioxide.

Upon addition to water, tris-buffer produces a pH of

9.2 to 9.8. It is desirable to adjust this pH to a lower value by the addition of hydrochloric acid or citric acid. Citric acid has the advantage of permitting pre-mixing of the buffer and acid. The amount of buffer added is determined by a number of variables. Mc-Farland and Norris' work was done with salt-water forms. The desired pH range for these forms is from 7.5 to 8.5. Since these values are within the range tolerated by fresh water forms, they are given here. For more limited buffering use 5 grams of tris-buffer per gallon plus 2.35 ml. of hydrochloric acid or 2.0 grams of citric acid. For a high buffer capacity use 15 grams of tris-buffer per gallon plus 6.40 ml. of hydrochloric acid or 14.11 grams of citric acid. The pH of the treated water should then be determined and if it is too high, i.e., above 8.5 it should be adjusted by the addition of more acid.

Sealed tanks appear to affect the amount of carbon dioxide dissolved in transport tank water. Haskell and Davies (1959) measured the effect of sealing a tank containing three pounds of trout per gallon. An hour after the tank was sealed the carbon dioxide reached a concentration of 31.3 p.p.m. When the tank was opened the carbon dioxide concentration dropped to 5.3 p.p.m. in one-half an hour.

Water cannot be hauled too satisfactorily in un-sealed tanks, but it is obvious that some means of removing stagnant air from above the water surface is desirable. In circulated tanks a venturi can be used to introduce air into the water, and an air escape pipe can be installed in the top of the tank. On non-circulated tanks that are transported on an open truck an arrangement such as shown in Figure 8 will assure flushing of the air over the water surface.

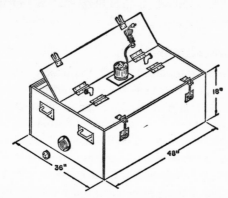

Specifications: *Constructed of either ¾″ exterior grade or marine plywood, all joints coated with waterproof resorcinol glue and nailed with 2″ brass boat nails. All surfaces of tank given two coats of expoxy swimming pool paint. Sealing flange ¾ x ¾″ white pine with ¼ x ¾″ sponge rubber gasket glued (3M adhesive) to upper surface after wood is painted. Drain consists of 1½″ pipe flange mounted over a 1½″ hole in end of tank. Air vents ½″ steel ells directed in opposite directions.*

8. *Hauling tank designed for a pick-up truck.*

Ammonia

Ammonia is a principal metabolic waste of fishes as well as being produced as a result of a break-down of feces and other organic waste. Ammonia is extremely toxic to fishes and probably can reach lethal levels during extreme crowding over extended periods of time. To date there has not been much done on the development of satisfactory materials for the adsorption of ammonia but starvation of fishes prior to transporting, and mechanical filtration are probably of value in reducing the production of ammonia.

Particulate Wastes

When transported, fishes frequently regurgitate and may also produce excess quantities of mucus. These materials plus fecal matter result in the water of transport tanks becoming badly fouled. As a result of bacterial decay the fouled water will exhibit a decrease in oxygen and an increase in carbon dioxide and ammonia. When fishes are to be crowded for a long period of time, some means of removing particulate wastes is desirable. Various mechanical strainers have been incorporated into water circulating systems. The system which shows the greatest promise is the diatomaceous earth filter used for filtering the water of swimming pools. Such a filter is capable of removing fine particles and is relatively easy to recharge (Norris, Brocado, Calandrino, and McFarland, 1960).

Use of Anesthetics in Transporting of Fishes

The metabolic rate of stimulated fishes is thought to be four to five times greater than that of the basal rate (Fry, 1957). By use of anesthetics it is possible to reduce the "activity" metabolic rate to the basal rate. Thus by use of anesthetics it is theoretically possible to increase the weight of fish hauled four to five times that which would be possible under normal conditions. McFarland (1960) is of the opinion, however, that due to the effects of an excess of pounds of fish per unit of water upon the build-up of waste products the practical increase in pounds of fish per unit of water is more in the order of two to three fold. He considers the most desirable stage of sedation to be that at which fish fail to respond to external stimuli but still retained equilibrium.

Fish should be treated prior to handling preparatory to transport. This, of course, presents a problem in cases where fish are to be collected directly from the wild and transferred to the transport unit. But some anesthetics that are not yet completely tested may be cheap enough to permit sedation of the fish prior to their being collected. Sodium cyanide may offer such a possibility. Of the anesthetics that he tested McFarland (1960) favors chloral hydrate, tertiary amyl alcohol, and methyparafynol (Dormison). The concentrations required to produce the desired state of sedation in the fishes McFarland used were 2 ml., 1.0 to 2.0 ml., and 3.0 to 3.5 gms. per gallon respectively.

Prophylactics

Various chemicals have been used to prevent out-breaks of disease and parasites resulting from the handling and transport of fishes. The general use of these various chemicals might be questioned inasmuch as any particular one is primarily suited to the control of only one group of pathogens while crowded fish may suffer epizootics from bacterial, fungal or protozoan pathogens. Where repeated difficulty is encountered with a particular pathogen, prophylactic treatment may be warranted. The bacteriostatic material most commonly used is acriflavin neutral at a concentration of 5 p.p.m. Water soluble terramycin at 20 p.p.m. will also suppress bacterial growth. To avoid outbreaks of protozoan and certain worm parasites during handling the author favors pre- and post-transport treatments with formalin. A thirty-minute dip in a 1 to 8,000 solution is recommended.

Transport Facilities

Facilities needed for hauling fishes vary greatly with the species and size of fish, with the distance they are to be hauled and the quantity involved. Polyethylene bags (4 to 6 mils) in cardboard boxes have come into general use for shipment of fish by common carrier. The units are usually made the maximum size accepted by the postal service but carry only four to five inches of water. The balance of the space is filled with oxygen. For occasional hauls, especially in cool weather, one pound of fish to three or four gallons of water can be transported limited distances without any special arrangements for aeration, etc. The containers should be a sealable type. Cooler boxes made of Styrafoam are very satisfactory, although this type transport facility can be greatly improved upon by supplying air with a hand-type automotive pump.

Next in order of cost and complexity of transport facilities is the sealed tank equipped with a 12-volt agitator. These units may be made in a variety of sizes and of various materials. For hauling in the trunk of an automobile, a tank two feet on each side and one foot deep is a convenient size. Such a tank may be made of heavy duty aluminum or ¾" plywood but in any event it should be equipped with a removable lid that rests upon flanges extending around the inside of the tank. A good seal is assured by use of a sponge rubber gasket on the flange, and clamps to hold the lid firmly in place. A hole is cut in the center of the lid to accommodate a 12-volt agitator. The agitator is energized by an extension cord wired to the automobile battery. For distances of 300 to 400 miles it will carry approximately one pound of fish per gallon. The unit described would have a capacity of approximately 30 gallons. At one pound of fish per gallon it would transport 27 pounds of fish. For transporting fishes up to 4 inches in length four perforated aluminum insert cans are used to facilitate handling the fish. When hauling larger fishes the unit can be used without the insert cans. For this type unit ice can be used for lowering or maintaining a low temperature. Anesthetics and buffer can also be used as described above.

A larger unit of this same type can be adapted to transport by a pick-up truck. The unit illustrated in Figure 8 has a capacity of approximately 135 gallons. Thus at one pound per gallon it has a capacity of approximately 120 pounds of fish. This tank may also be equipped with insert cans. Such a unit is large enough to warrant filtration which would be possible with the addition of a self-priming, centrifugal pump with a suction line attached to the drain of the tank and spray jets installed in the top of the tank. A swimming pool type filter could be installed in the pressure line. It would still be desirable to retain the 12-volt agitator, however.

APPENDIX

LITERATURE CITED

INDEX

I. *Dosage calculations.*

For a 1 percent solution, add:

> 38 grams per gallon
> 1.3 ounces per gallon
> 10 grams per 1000 ml.
> 38 ml. per gallon
> 10 ml. per 1000 ml. (1 liter)

For other percent solutions, multiply by factors concerned. Thus, a 5 percent solution is 5 × 38 grams per gallon.

II. *Feeding drugs.*

To feed a 1 percent level in food, add:

> 5.4 grams per lb. of food
> 0.2 ounces per lb. of food
> 83 grains per lb. of food

For other dosage levels, multiply by appropriate factor.

EXAMPLE: For a 2 percent diet level, multiply 5.4 grams by 2 and add to one pound of food.

For a 0.2 percent level, multiply by 0.2.

III. *Conversion Chart* (Part I)

Unit	Gallon	Quart	Pint	Pound	Ounce	Fluid Ounce
1 gal.	1.0	4.0	8.0	8.345	133.52	128.0
1 qt.	0.25	1.0	2.0	2.086	33.38	32.0
1 pt.	0.125	0.5	1.0	1.043	16.69	16.0
1 lb.	0.12	0.48	0.96	1.0	16.0	15.35
1 oz.	0.0075	0.03	0.06	0.0625	1.0	0.96
1 fl. oz.	0.0078	0.031	0.062	0.062	1.04	1.0
1 cu. in.	0.0043	0.017	0.035	0.036	0.573	0.554
1 cu. ft.	7.481	29.922	59.848	62.428	998.848	957.48
1 cc.	0.0003	0.001	0.002	0.002	0.035	0.034
1 liter	0.264	1.057	2.1134	2.205	35.28	33.815
1 gram	—		0.002	0.002	0.0353	0.034

IV. *Conversion Chart* (Part II)

Unit	Cubic Inch	Cubic Foot	Milliliter	Liter	Gram
1 gal.	231.0	0.1337	3785.4	3.785	2785.4
1 qt.	57.749	0.0334	946.36	0.95	946.35
1 pt.	28.875	0.0167	473.18	0.47	473.18
1 lb.	27.67	0.016	453.59	0.454	453.59
1 oz.	1.73	0.001	28.3	0.03	28.35
1 fl. oz.	1.8	—	29.57	0.03	29.41
1 cu. in.	1.0	0.0006	16.39	0.0164	16.3
1 cu. ft.	1728.0	1.0	28322.0	28.316	28318.58
1 ml.	0.061	—	1.0	0.001	1.0
1 liter	61.025	0.0353	1000.0	1.0	1000.0

V. *Gravimetric and Volumetric Equivalents.*

Unit of Measure	Equivalent
1 milligram per liter	1 part per million
1 kilogram	2.205 pounds
1 pound	453.6 grams
1 grain per gallon	17.12 parts per million

Unit of Measure	Equivalent
1 grain per gallon	142.9 pounds per million gals.
1 part per million	0.0584 grains per gallon
1 gallon	231 cubic inches
1 cubic foot	7.48 gallons
1 cubic foot of water	62.4 pounds
1 gallon of water	8.34 pounds
1 gallon	3.785 liters
1 liter	0.2642 gallon
1 liter	1.057 quarts
1 liter	61.02 cubic inches
1 inch	2.54 centimeters
1 centimeter	0.3937 inch
1 part per million	8.34 lbs. per million gals.
1 pound per million gallons	0.1199 parts per million
1 gram	15.432 grains
1 pound	7000 grains
1 meter	39.37 inches
1 cubic centimeter	0.0610 cubic inch
1 cubic inch	16.387 cubic centimeters
1 quart	0.046 liter
1 gram	0.0353 ounce
1 ounce	28.3495 grams

VI. *Temperature Equivalents.*

Degrees Centigrade = $\frac{5}{9} \times$ Degrees Fahrenheit − 32.

Degrees Fahrenheit = $\frac{9}{5} \times$ Degrees Centigrade + 32.

VII. *Dissolve Gases Equivalents.*

Oxygen in p.p.m. $\times$ 0.7 = c.c., or ml. per liter.

Oxygen in c.c. or ml. per liter $\times$ 1.429 = Oxygen in p.p.m.

Carbon Dioxide in p.p.m. $\times$ 0.509 = c.c. or ml. per liter.

Carbon Dioxide in c.c. or ml. per liter $\times$ 1.964 = Carbon Dioxide in p.p.m.

VIII. *Grams per Liter to Ounces per Gallon.*

Grams per Liter	Ounces (Avoir.) per Gal. (U.S.)	Percent Solution
100	13.35	10
90	12.02	9
80	10.68	8
75	10.01	7.5
70	9.35	7
60	8.01	6
50	6.68	5
40	5.34	4
30	4.01	3
25	3.34	2.5
20	2.67	2
10	1.34	1
9	1.20	0.9
8	1.07	0.8
7	0.93	0.7
6	0.80	0.6
5	0.67	0.5
4	0.53	0.4
3	0.40	0.3
2	0.27	0.2
1	0.13	0.1
0.75	0.10	0.075
0.5	0.07	0.05
0.25	0.03	0.025

IX. Weights (in grams) of Chemicals Required to Produce Desired Dilutions in Known Volumes of Water.

Dilution	Gallons of Water					
	1	5	10	15	20	25
1:1,000	3.78	18.93	37.85	56.78	75.70	94.60
1:2,000	1.89	9.46	18.93	28.39	37.85	47.30
1:3,000	1.26	6.31	12.62	18.92	25.23	31.50
1:4,000	0.95	4.73	9.46	14.19	18.92	23.60
1:5,000	0.76	3.79	7.57	11.35	15.14	18.90
1:10,000	0.38	1.89	3.79	5.68	7.57	9.40
1:15,000	0.25	1.26	2.52	3.78	5.05	6.30
1:20,000	0.19	0.95	1.89	2.84	3.78	4.70
1:100,000	0.038	0.19	0.38	0.57	0.76	0.90

NOTE: The table can be used to obtain any dilution for any volume. For example, if a 1:40,000 dilution is desired in 50 gallons of water, use the figure for the 1:20,000 dilution for 25 gallons (4.70).

Tables selected from Hutchens and Nord 1953 *Fish Cultural Manual*, mimeo., 220 pp.

LITERATURE CITED

American Public Health Association, *et. al.*
  1955. Standard Methods for the Examination of Water, Sewage, and Industrial Wastes, 10th Ed. (New York, 1955). 522 pp.

ATZ, JAMES W. and GRACE E. PICKFORD.
  1959. The use of pituitary hormones in fish culture. Endeavour XVIII (71):125–129.

BREDER, C. M., JR.
  1931. On organic equilibria. Copeia, 1931(2):66.

BRIDGES, W. R.
  1958. Sodium cyanide as a fish poison. Special Scientific Report, Fish. No. 253. U. S. Fish and Wildlife Service, 11 pp.

BROWN, MARGARET E.
  1957. The Physiology of Fishes, Vol. 1. Metabolism, 447 pp. and Vol. 2. Behavior, 526 pp. Academic Press, Inc. (New York, 1957).

BURROWS, ROGER E.
  1949. Prophylactic treatment for the control of fungus (*Saprolegnia parasitica*) of salmon eggs. Prog. Fish-Cult., 11:97–103.

DAVIS, H. S.
  1956. Culture and Disease of Game Fishes. Univ. of Calif. Press. (Berkeley and Los Angeles, 1956). 332 pp.

FISH, FREDERIC F.
  1941. Notes on *Costia necatrix*. Trans. Amer. Fish. Soc., 70:441–445.

FISH, FREDERIC F.
1942. The anaesthesia of fish by high carbon dioxide concentrations. Trans. Amer. Fish. Soc., 72:25–29.

FRY, F. E. J.
1957. Aquatic Respiration of Fishes, Chapter 1 in Brown's Physiology of Fishes, Vol. 1. Metabolism. Academic Press, Inc. (New York, 1957). 447 pp.

GRIFFIN, PHILLIP J.
1953. The nature of bacteria pathogenic to fish. Trans. American Fish. Soc., 83:241–253.

GORDON, MYRON.
1950. Fishes as Laboratory Animals. Chapter 14 in Farris' The Care and Breeding of Laboratory Animals. John Wiley and Sons, Inc. (New York, 1950). 515 pp.

HASKELL, DAVID C. and RICHARD O. DAVIES.
1959. Tank cover limits capacity of trout transportation tanks. Prog. Fish. Cult., 21(4):187.

INNES, WILLIAM T.
1951. Exotic aquarium fishes. 12th Ed. Innes Publishing Co. (Philadelphia, 1951). 521 pp.

JACOBS, D. L.
1946. A new parasitic dinoflagellate from freshwater fish. Trans. Amer. Microb. Soc. 65:1–17.

KAWAMOTO, N. Y.
1961. The influence of excretory substances of fishes on their own growth. Prog. Fish. Cult. 23(2):70–75.

McFARLAND, WILLIAM N. and KENNETH S. NORRIS.
1958. The control of pH by buffers in fish transport. Calif. Fish and Game, 44(4):291–310.

McFARLAND, WILLIAM N.
1960. The use of anesthetics for handling and transport of fishes. Calif. Fish and Game, 40(4):407–431.

MUENCH, BRUCE.
  1958. Quinaldine, a new anesthetic for fish. Prog. Fish.
        Cult., 20:42–44.

NIGRELLI, ROSS F.
  1953. The fish in biological research. Trans. of the N. Y.
        Acad. of Sci. Ser. II, V.15:183–186.

NORRIS, KENNETH S., et. al.
  1960. A survey of fish transportation methods and equip-
        ment. Calif. Fish. and Game, 46(1):5–33.

PERLMUTTER, ALFRED, and EDWARD WHITE.
  1962. Lethal effect of fluorescent light on the eggs of
        brook trout. Prog. Fish. Cult., 24(1):26–30.

PIPER, ROBERT G.
  1958. Ulcer disease of trout. Fish and Wildlife Service
        Fishery Leaflet 466. 3 pp.

ROSE, S. MERYL.
  1959. Failure of survival of slowly growing members of
        a population. Science, 129(3355):1026.

RUCKER, ROBERT R., et. al.
  1953. Infectious diseases of pacific salmon. Trans. Amer.
        Fish. Soc., 83:297–412. (1953).

SNEED, KERMIT E., and H. P. CLEMENS.
  1959. The use of human chorionic gonadotrophin to
        spawn warm-water fishes. Prog. Fish. Cult., 21:117–
        120.

SNIESZKO, S. F.
  1954. Therapy of bacterial fish diseases. Trans. Amer.
        Fish. Soc., 83:313–330.

SNIESZKO, S. F.
  1958a. Fish furunculosus. Fish and Wildlife Service Fish-
        ery Leaflet 467.

SNIESZKO, S. F.
  1958b. Columnaris disease of fishes. Fish and Wildlife
        Service Fishery Leaflet 461. 3 pp.

STRAWN, KIRK and CLARK HUBBS.
  1956. Observations on stripping small fishes for experi-
    mental purposes. Copeia., 2:114–116. (May, 1956).

SWINGLE, H. S.
  1956. Appraisal of methods of fish population study, Part
    IV. Determination of balance in farm fish ponds.
    Trans. North American Wildlife Conference,
    21:298–322.

VAN DUIJN, C. JR.
  1956. Diseases of Fishes. Water Life, Dorset House,
    Stamford Street, London.

WATSON, STANLEY W.
  1954. Virus diseases of fish. Trans. Amer. Fish. Soc.,
    83:331–341.

WELCH, PAUL.
  1948. Limnological Methods. Blakiston Company (Phila-
    delphia and Toronto, 1948). 581 pp.

INDEX